SCIENTIFIC BENGAL

SCIENTIFIC BENGAL
Science, Technology, Medicine and Environment under the Raj

Chittabrata Palit

Publications

KALPAZ PUBLICATIONS

DELHI-110052

Scientific Bengal

Rs. 750

ISBN : 81-7835-496-9

Published in 2006 in India by
Kalpaz Publications,
C-30, Satyawati Nagar, Delhi-110052
E-mail : kalpaz@hotmail.com

Laser Typesetting at : Master Computers, Delhi
Printed at : Salasar Imaging System, Delhi

Dedicated to

Subhabrata Palit

Contents

SECTION-IV : ENVIRONMENT

Preface

The Essays collected in this volume are all papers presented to national and international conferences in recent years. Some of them have also been printed in the proceedings volumes. It is not possible to thank the sponsors individually for permission to print them in this book. This may be treated as a general acknowledgement.

Social History of Science has now acquired the status of mainstream history as it charts material progress of a nation. Though the present essays deal with development in Bengal, they have application at other regional levels and national level. Bengal was the cradle of the first flush of scientific pursuits in the colonial era which was later nationalised in all parts of India. Its appeal as a whole is not only national but also universal for parallel studies in developing societies. It has got four sections: Science, Technology, Medicine and Environment. No work has so far appeared including all four aspects of scientific development in a region within two covers. This can claim to be a trend-setter.

The author's labours will be rewarded if the book is well received by the reading public.

In the end, it must be mentioned that research for this work was liberally funded by Wellcome Trust, London, U.K., CIES/ Fulbright Fellowship Fund, Washington, U.S.A., NISTADS, New Delhi and INSA, New Delhi.

Chittabrata Palit

1

Introduction

History of Science has become a raging theme in the post-modernist inter disciplinary research. So far this field was monopolised by scientists in India. They have generally written about disciplinary developments of various sciences in India as in S.N. Sen et al authored. A concise history of the social context of the origin and development of these sciences is not so much available in their writings. Historians, therefore, have attempted to write the social history of sciences mainly explaining the social context and content of the rise of sciences in India.

The present collection of essays has been written from the historian's point of view for a specific region i.e. Bengal, which was the cradle of scientific pursuits in India. This micro-study should provide more depth than macro-studies attempted by A.K. Biswas, Ashis Nandy and Deepak Kumar.

The essays have been divided into four sections i.e. Science, Technology, Medicine and Environment, altogether containing fifteen essays. Section A on science contains three essays. The first is on the origin of the Asiatic Society and the role of Willam Jones in promoting scientific studies in it. The Asiatic Society has been shown as the brain trust of the Company's government for collection of information on Indian resources and their application for administrative and commercial requirements. In this sense, it was the beginning of colonial science in India and William Jones was its father figure. The next essay is on Serampore Missionaries and their contribution to colonial science. Willam Carey, Felix Carey and John Mack contributed to printing and paper technology, botany, physiology and chemistry for evangelical and imperial purposes not always clear to the worshipful intelligensia. Carey has been specially highlighted in this essay. The third essay in this section is on the dawn of national science in response to colonial science. The denial of scientific education to Indians have been pointed out by Rammohan Ray, Akshay Kumar Dutta, Rajendralal Mitra and others. But the doyen of a national science association was Dr. Mahendralal Sircar who after a decade of struggle was able to establish the Indian Association for the Cultivation of Science in 1876. His life and work has been detailed in this essay. The legacy of IACS was carried by the Dawn Society (1902) and the National Council of Education (1906).

Section B on technology begins with an essay on Sitanath Ghosh, the pioneer of the swadeshi technology in Bengal. This obscure personality has been resurrected from the pages of history by the author for his indigenous inventions of various instruments of which the most famous was the Magnetic Healer. He at least gets his rightful place in history in this essay. This is followed by an institutional history of swadeshi technical education in which Bengal Technical Institute, the first proper institute of technology has been studied in detail from its inception in 1906 to its merger with the College of Engineering and Technology in 1928. The last essay in this section is on the National Council of Education which was the parent of both BTI and CET. The thrust of this essay is however, on the CET which had an illustrious career from 1928-1947. This was the golden period of swadeshi technical education in Bengal and produced nearly five hundred engineering graduates by 1947 who were responsible for industrialisation and modernisation in India.

Section C is on medicine. The first essay in this section sketches the origin and growth of western medicine in Calcutta in the nineteenth century with reference to both hospitals and medical education in general. The centre-piece is of course the Calcutta Medical College which played a pivotal role in the spread of western medicine in India. The next essay studies the indigenous system of medicine in colonial Bengal namely Ayurvedic, its decline, survival and revival. This is a path-breaking essay and vastly improves upon what Brahmananda Gupta wrote on the subject in

Charles Leslie's Asian Medical Systems. The third essay is an exhaustive treatment of colonial public health policy which caused epidemics. It also elaborately deals with popular response to the public health policy and the epidemics. This is perhaps the first Indian reply to western interpretations of public health in India as found in David Arnold, Ira Klein and Mark Harrison. The last essay in this section is on the life and work of Ronald Ross. It shows how even a British medical scientist was thwarted in his research work on Malaria while he was serving as an IMS doctor under the colonial government. His discovery of the mosquito cycle of malaria earned him the Nobel Prize in medicine in 1902 but his programme of sanitation for India was scotched and this friend of India was sent back home for this Philanthropic work.

Section D deals with environment. The opening essay in this section is the British penetration in to the North-Eastern hills and the creation of Darjeeling as a hill station and summer capital of Bengal around 1840. This was the first major example of tapping natural resources of the hills for colonial extraction. Tea plantation was raised here for the first time for commercial exploitation and this set the environment in adverse motion. The second essay is also on tapping of the hinterland by the colonial authorities. This deals with the uprooting of the Santals from the uplands of Chhotanagpur and their fanning out in different parts of Bengal. Here also the environment was a casualty and the Santal village community in disarray had to face the problem of broken villages and peasantisation in the Bengal plains. No wonder that they rose in revolt many a time. The modern concept of social forestry would have rehabilitated them in the nineteenth century. They still suffer from vagrancy and there is still scope for the application of the norms of social forestry for them even today. The following essay is an overview of green imperialism as placed by Richard Grove and contested by Gadgil and Guha. It led ultimately to loss of ecological balance. Free India has also not taken the lesson of history and continues to plunder nature. The last essay is on the thoughts of Rabindra Nath Tagore and Gandhi on environment. Tagore was aware of the plunder of villages for the embellishment of cities which cause ecological imbalance. He sought out ways and means to eradicate it by various ways and means, the most important of which was an embargo on the incessant demands on nature and symbiosis with it. Gandhi thought alike.

All told, the collection of essays in this volume promises to be a panorama of scientific developments in Bengal for which there is not a single book available so far. This study of the region can be a trend-setter for any other region of India and macro study for the sub-continent. This is a major exercise in the history of sciences, a new and fertile area of post-modernist research. This book should find its place in the growing shelf on the subject in any library.

SECTION-I : SCIENCE

2

Orientalism, William Jones and Science at the Asiatic Society

Orientalism

Orientalism, in the ordinary sense, meant oriental studies by William Jones, Halhed, Colebrooke, Wilkins, Prinsep etal connected with the activities of the Asiatic Society of Bengal founded in 1784 and patronised by that Persian knowing first Governor General of Bengal, Warren Hastings. It also meant the continuity of classical learning at the Sanskrit College at Calcutta and Benares and Persian-Arabic studies at the Calcutta Madrasa. Orientalism figured again during the Anglicist debate over the introduction of a new educational policy during the time of Lord William Bentinck in 1835 in which the Orientalists, Shakespeare Prinsep, H.T. and James, Elliot, MacNaaghten and Sutherland fought a losing battle against the Anglicists like Macaulay. This narrativistic account of Orientalism in the literal sense continued unabated without any serious attempt at any interpretation of the discourse till 1964 as can be seen in Cannon's *Oriental Jones* (1964). David Kopf (1969) though writing after S.N. Mukherjee's first interpretative work on the subject (1968) suffered from the usual faults of a researcher trapped by the monism of his own thesis. I quote : The Orientalists served as avenues linking the regional elite with the dynamic civilization of contemporary Europe. They contributed to the formation of a new Indian middle class and assisted in the professionalization of the Bengali intelligentsia. They started schools, systematized, languages, brought printing and publishing to India and encouraged the proliferation of books, journals, newspapers and other media of communication. Their impact was urban and secular. They built the first modern scientific laboratory in India and taught European medicine. They were neither static classicists nor averse to the idea of progress and they both historicized the Indian past and stimulated a consciousness of history in the Indian intellectual. It was they who transmitted a new sense of identity of Bengalis.[1]

The orientalists had no role in the creation of the Bengali middle class. He has virtually summarized the intellectual impact of British rule as a whole unilaterally in the language of the ruling class and treated the Bengali intelligentsia as dead metal. Much of the work credited to the orientalists were actually accomplished by the Anglicists though Bengalis like Rammohan and the Pandits and Munshis of Fort William College had their share in it. Only the construction of the historical past, however faultily and Eurocentrically, was their work aided by the Pandits. Kopf's work is but an instance of 'orientalism' already stated as an attitude by S.N. Mukherjee (1969) but more powerfully by Edward Said in 1978 in his 'Orientalism' as a theoretical discourse. But he, in turn, owed his ideas to Michel Foucault. Though not directly but tacitly, Michel Foucault in 'The Order of Things' (1970) and 'The Archaeology of Knowledge' (1972) placed Orientalism in a problematic discourse. Foucault wrote in 1970: all knowledge is rooted in a life, society and a language that have a history; and it is in that history that knowledge finds the element enabling it to communicate with other forms of life, other types of society, other significations'.[2] He wrote more explicitly in 1972: 'Whatever their formal level may be, I shall call these themes and theories strategies. The capacity to invest this discourse in decisions, institutions or practices is in fact confined (sometimes with the addition of legal sanctions) to a particular group of individuals in the bourgeois societies that we have known since the sixteenth century. There is no knowledge without a particular discursive practice and my discursive practice may be defined by the knowledge that it fosters.'[3] Taking the cue from Foucault and perhaps from Antonio Gramsci's theory of cultural hegemony[4], E.H. Said hit the iron on the head when he wrote: 'It is rather a distribution of geopolitical awareness into aesthetic, scholarly, economic, sociological, historical and philological texts.., it is .. a certain will or intention to understand, in some cases to control, manipulate, even to incorporate what is manifestly a different or alternative or novel world.'[5] Orientalism as Said defines, was concerned with 'an irresistible impulse always to codify, to subdue, the infinite variety of the Orient to a complete digest of laws, figures, customs and works.'[6] Divide et Empera, the imperialist motto meant classify and control. Though Said applied his history to the Anglo-French colonial domains of the Middle East and Africa with reference to Arabic-Persian linguistics and Islam, the logic applies equally codgently in Indology. Essentially, it is the Hegelian idea of world history and philosophy of history. Orient is viewed as the binary opposite of the Occident, the 'Other' whose intellectual tradition must be learnt, classified, controlled and subsumed under western intellectual hegemony. To the British rulers, colonial India was the 'Other' which had to be thoroughly digested in its religion, culture, language, law, geography, science and medicine as things in a state of nature essentialising the stable factors or constants of an Asiatic Society and minimising the variables so as to be able to put the information to a system analysis as in a computer in order to discipline such societies and rule them effectively and appropriate its natural and human resources. Even Marx could not escape such a Eurocentric, orientalist view in describing the 'Other' as the Asiatic mode of production which the West must destroy and build the basis for capitalism.[7] S.N. Mukherjee in his study of William Jones (1968) for the first time made it an analytical work on the 18th century British attitudes to India. He at least viewed the proceedings of the Asiatic Society as European discoveries of India and an effort to affiliate and appropriate. Indian classical lore and even natural phenomena within the European corpus of knowledge. He came close to Said's thesis but in his introduction to the second edition of the same book (1987) refers to the Orientalism controversy in passing without really revising his work as he believes that Said deals with 'popular orientalism' (mainly a literary style) and confuses it with scholarly orientalism.[8] This is arbitrary and unacceptable as Said's Orientalism takes place when one set of hegemonic texts accosts another set and tries to subsume it which is what happened in India in the name of comparative studies. Bernard Cohn characterizes the discovery of India's past as a 'European project, the end being to construct a history of the relationship between India and the West, to classify and order and locate their civilizations on an evaluative scale of progress and decay.'[9] Ronald Inden sees in Orientalism the tendency to perpetuate the East-West binary contrasts by essentialising and representing both in stereotypes of spiritualism, other-worldiness, mysticism, caste, changeless village community and oriental despotism on the one hand and individualism, liberty, democracy, capitalism, welfare etc. on the other. All attibutes are internationally changing in all three worlds and an international discourse is emerging on human race as a whole. But this decentres only the orientalist construct of Indian atrophy and dovetails it into the western universe of progress. The quintessence of Indian culture lies in its rich repertoire of philosophy, religion, literature, art, music, and other performing arts as also mathematics, astronomy, physics, alchemy and medical lore which constitute its autonomy.[10]

William Jones 1747-1794

William Jones became an Oxford graduate with a reputation for his command over Greek, Latin, Arabic and Persian languages. But this kind of proficiency did not secure him any job except that of private tutor in a few lordly families. He got the patronage of Shelburne, Shipley and Ashburton and the Duchess of Devonshire but no good job was yet in sight. He fought unsuccessfully for a parliamentary seat from Oxford in 1780. He was prepared to go to America, the land of liberty and saw Benjamin Franklin at Paris. At home, he studied law for a career but it did not fetch him any better job than Commissioner of bankruptcy carrying less than £100 as salary with which it was impossible to settle in London with family. He was not a successful barrister and because of his feeble health, even his fiancee, Anna Maria despaired of his success as a lawyer. In such a desperate situation, he turned to radical politics and wrote the pamphlet, 'the Principles of Government in a Dialogue between a Scholar and a peasant', for the Society for

Constitutional Information in which he recommended the peasant to be armed for justice. It was a libel.[11] But writing to the Lord Chancellor Thurlow, he swore: 'As to my politics which he has heard much misrepresented, his Lordship may be assured that I am no more a republican than a Mahomedan or Gentoo and that I have ever formed my opinion from what appeared to me on the calmest enquiry, the true spirit of our constitution.'[12] This was in connection with his solicitation for a judgeship on the Supreme Court of Bengal which he had been seeking for some time past through his friends and patrons, Kenyon, Bruke, Shelburne and Ashburton. This reflects both his opportunism and early traits of orientalism. In 1781-82, he gave assistance to Burke on Indian legislation and Ashburton refers to 'the important stake still left in India or your particular interest in it in which I consider that of the public so materially involved.'[13] Even Anna Maria mentions in her letter to Georgina about 'the India scheme or plan' which was something more than a mere judgeship.[14] Ashburton and Kenyon lobbied and Shelburne finally recommended him for the judgeship to George III who approved Jones for the post for his competence as a lawyer and his knowledge of the eastern languages.' This was a high-minded royal appointment for the post of a puisne judge under the East India Company. More unusual was his instant knighthood on request by Jones himself on 20th March 1783. Though his India scheme or stake has not been deciphered by Jone's biographers, one can infer that he had a very special mission in the role of a puny judge.[14] It must have been to gather all kinds of information on India for the empire and the West as a whole to assimilate but more immediately to produce a digest of Indian Law to complete which he became a martyr for this knowledge was power to rule India effectively. To quote Warren Hastings, 'Every accumulation of knowledge and especially such as is obtained by social communication with people over whom we exercise dominion founded on the right of conquest is useful to the state...it attracts and conciliates distant affections; it lessens the weight of the chain by which the natives are held in subjection and it imprints on the hearts of our countrymen the sense of obligation and benevolence.'[15]

The Asiatic Society

Jones boarded the frigate crocodile for India on 12th April 1783. He drew up a memorandum for his future activities in India. It is worth quoting here: The Objects of enquiry during my residence in Asia:

1. The laws of the Hindus and Mohammedans.
2. The history of the ancient world.
3. Proofs and illustrations of Scripture.
4. Traditions concerning the deluge etc.
5. Modern Politics and geography of Hindustan.
6. Best mode of governing Bengal.

7. Arithmetic and geometry and mixed sciences of Asiaticks.

8. Medicine, chemistry, surgery and anatomy of the Indians.

9. Natural products of India.

10. Poetry, rhetoric and morality of Asia.

11. Music of the Eastern Nations.

12. The She-King or 300 Chinese Odes.

13. The best accounts of Tibet and Kashmir.

14. Trade, manufactures, agriculture and commerce of India.

15. Mughal Constitution.

16. Maharatta Constitution.

S.N. Mukherjee adds: 'Bur he knew that he had to gather rare and hitherto unknown manuscripts and seek their explanation from the natives and this would involve endless discussions with Asians on religion, history, art and other subjects.[16] This is virtually the blueprint of the Asiatic Society to be founded shortly. Still Mukherjee thinks that he had not yet thought of a plan to form a society to investigate into the science, history and arts of Asia. There was already a nucleus in Bengal formed by a band of young British civilians interested in Asiatic studies under the patronage of the Governor General Warren Hastings. There were Charles Wilkins, Nathaniel Halhed, John Shore, Francis Gladwin, John Carnac, Jonathan Duncan and William Chambers who had already produced digests of some of the standard works on law and religion in India. Wilkins had translated Srimadbhagabadgita and started work on Mahabharata. Halhed had produced his famous Gentoo Laws and a Grammar of the Bengali Language. Halhed and Shore collected many Persian manuscripts which were translations of original Sanskrit texts. Hastings had his vision of ruling India by mild government in this precarious and formative period of the empire. Hastings has already been quoted on this point. He founded the Calcutta Madrasa and the Sanskrit Colleges at Benares and Calcutta, 'to soften the prejudices excited by the rapid growth of the British dominions.' Halhed put it more succinctly: The importance of the commerce of India and the advantages of a territorial establisment in Bengal have at length awakened the attention of the British legislature to every circumstance that may conciliate the affections of the natives or ensure stability to the acquisition. Nothing can so favourably conduce to these two points as a well timed toleration in matters of religion and adoption of such original institutes of the country as do not immediately clash with the laws or interests of the conquerors.[17] But it will be fatuous to say that Hastings patronised oriental studies only for the purpose of real politik. Mukherjee justly writes: 'He came to India at an early age and a long stay in the country had made him an admirer of Indian manners and customs. He mastered the Persian language, gathered Indian paintings and manuscripts and in his letters to his wife, he used to quote from the Gita which he found a source of inspiration.[18] This is the other face of orientalism by which the Europeans wanted to enrich their own culture. Jones was destined to carry out the double mission. Jones's appointment as puisne judge was just to provide him with the purse to sustain his more worthy purpose to knead together the disintegrating school of orientalists as Shore avers, 'if Mr. Jones should, as we are taught to expect, arrive in Bengal.'[19] Jones arrived at Calcutta on 25th September 1783 and joined his judgeship in December. In January 1784, he sent out a circular letter giving a plan to establish a society to encourage oriental studies in Calcutta. Thirty gentlemen responded to the letter and on 15th January they gathered in the Grand Jury Room of the Supreme Court of Calcutta. The Chief Justice, Sir Robert Chambers, an old Fellow of University College, Oxford presided over the first meeting where Jones delivered his first discourse outlining the purpose of the proposed society. Its objects of enquiry would be man and nature, 'whatever is performed by the one and produced by the other' in Asia. In the Second discourse, he expanded on his idea: 'We ought not to condem the people of Asia from where researches into nature, works of art and inventions of fancy, many valuable hints may be derived for our own improvement and advantage. If that indeed were not the principal object of our institution, little else could arise from it.'[20] In the essay on the Poetry of Eastern nations, he further explained, 'I cannot but think that our European poetry has subsisted for long on the perpetual repetitions of the same images and incessant allusions to the same fables.'[21] Asiatic Society was to launch European discovery of India for their own appropriations. This was the kind of orientalism Said talks about and Cohn interprets as western invasion into Indian episteme, the western construct not necessarily representing the true meaning of the Indian text and trying to affiliate it to the western system of thought by the so-called comparative studies. This was Jones's higher calling for which he was specially brought to India and not merely to inform a brutish British administration for benign rule. It was all set before he even set foot on Calcutta or else one cannot explain the rapidity with which Asiatic Society was launched within three months of his arrival. Such a body had to be non official, informal and autonomous though the administration backed it from the inception. It was like the Batavian Society founded by the Dutch in the East Indies in 1778. Warren Hastings rightly refused to become its President in order not to reduce it to a government department of information,though he helped it in every way. On his refusal, Jones was made the President on 5th February 1784 in the fitness of things. The later Governor Generals, Shore and Cornwallis attended all its sessions and sanctioned funds for the printing of its proceedings and feeing the Indian Pandits engaged in compiling the Law Digests. All the founder-members belonged to the officialdom but Indians were kept out of it till 1829. Jones himself advised against it though translations of their work by this band of officials was to be encouraged and some of their original compositions were to be printed in the journal. O.P. Kejariwal tries to dissociate Asiatic Society from a direct link with the government to prove British altruistic motive. But as Abhijit Mukherjee has shown amply, such a defence does not bear scrutiny.[22] It was a charmed circle of officials already engaged in oriental studies. The attendance was fifteen on an average. Asiatic Researches volume 1 embodying the discourses at the Asiatic Society was first published in 1788. From 1787 till his death in 1794, Jones gave all the annual discourses himself and most of these papers formed the corpus of the Researches in its first four volumes. Avid for learning Sanskrit, he purchased a house in Krishnagar in 1785, his third university and appointed Ramlochan Sharma as his Sanskrit tutor. Pandit Radhakanta Sharma who worked with Hastings and Shore was his research adviser. During his visit to Benares in 1784, he had met Ali Ibrahim Khan, an Indian scholar appointed chief magistrate of Benares in 1782 who later contributed articles on Hindustani poets and Trial by Ordeal for the Asiatic Researches. Ghulam Hussain, the historian also became his friend. Jagannath Tarkapan-chanan, the scholar extraordinary of that age complied and edited the Digest of Hindu laws for Jones with his own commentaries which was to supersede Halhed's Gentoo

Laws. Abhijit Mukherjee in his remarkable essay on 'European Jones and Asiatic Pandits' discusses in details the collaborative efforts of these Pandits in promoting oriental studies.[23] There was no opposition from them to imparting the language of the Gods to the foreigners as made out by Cannon.[24] Most of the eleven pandits whom Hastings appointed to compile Vivadarnavasetu or a Code of Gentoo Laws in 1773 cheerfully cooperated at a pittance. Hastings had the following to say about them in the introduction to the book: 'The most learned Pandits of the province were desired to assist in this work. They readily and cheerfully gave their consent and without requiring but as I recollect even refusing any pecuniary gratification, they undertook and executed a complete abstract of their laws.'[25] Five of them were still living in 1788 and one of them, Pandit Gourikanta corrected Jones's copy of the Vivadarnavasetu. Jones also met Kashinath in Benares, the Pandit who gave scholarly assistance to Wilkins and sent Govardhana Kaul to assist Jones in preparing the Digest of Hindu Laws and also contributed to the first volume of the Asiatic Researches. Jones who had no knowledge of Sanskrit on his arrival in September 1783 picked up the language in two years well enough to compose verses in it and speak it fluently. Sarvaru Tewari from Behar was another Pandit who assisted Jones in preparing the Digest. Ramlochan's ability is reflected in his deciphering of the six inscriptions sent by Gen. Carnac from the Fort of Tanna, the capital of Salset in February, 1787 which was published in the Asiatic Researches. He and Govardhana Kaul assisted Jones in reconstructing the chronology of the Hindus. Radhakanta Sharma was a consummate scholar. He wrote the 'Puranarthaprakasha' which Jones received from Shore and hailed as a treasure. Radhakanta remained his associate from 1787 till his death in 1794. Jones borrowed heavily from Radhakanta's work on the king-list in the Puranas in his paper.' On the Chronology of the Hindus' published in the Asiatic Researches.[26] Radhakanta also read the Delhi-Topra Pillar inscription of the Chahamana king Visaladeva of 11th century A.D. which was duly published in the Asiatic Researches.[27] The Digest of Hindu and Mohammedan Laws superintended by Jones from 1788 was the handiwork of Radhakanta and Jagannath Tarkapanchanan, the venerable sage of Tribeni. In March 1794, Radhakanta was 91 and still found indispensable by Jones for completion of his Digest. But Jones predeceased him. Jagannath was not Jones's Pandit. He was venerated as the great sage throughout upper India. He was appointed for the Code in 1788 on a pay of Rs. 300 a month and Rs. 100 for his assistant. In February 1702, he submitted his work Vivadangarnava' or The Sea of Controversial Waves. Colebrooke in his preface to the Digest wrote in appreciation: 'the venerable author, Jagannath has added a copious commentary, sometimes indeed pursuing frivolous disquisitions, but always fully explaining the various interpretations of which the text is susceptible.[28] Radhacharan, the chief pandit of Samuel Davis at Benares helped him to write a paper 'On the Astronomical Computations of the Hindus.'[29] He also supplied a glossary of scientific terms to go with it and presented to Jones a copy of Surya Siddhanta on the basis of which Jones wrote his 'A Supplementary Essay on Indian Chronology'.[30] Radhakanta's work on the Puranas mentions Nanda and Maurya King Chandra Gupta which helped Jones to identify the Greek corruption of the name of the Indian king Sandroxottas to be Chandra Gupta. Thus much of what Jones accomplished was due to the spadework of the Indian Orientalists pejoratively called Pandits or scribes. Jones was only their research coordinator. Even his fame as the father of comparative philology which rests on his paper, 'On the Hindus' was composed when his know-ledge of Sanskrit was only four months old and naturally imperfect. But we are still carried away: 'The Sanskrit language whatever be its antiquity is of a wonderful structure; more perfect than the Greek, more copious than the Latin and more exquisitely refined than either yet bearing to both of them a stronger affinity....'[31] What he was doing was already attempted by Boxhorn etc. i.e., to affiliate Sanskrit into an Indo-European family of Scythian origin. This is the most sophisticated form of Orientalism as we understand it today thanks to Said and Cohn. Add to this the pioneeing observations of William Jones when he wrote that 'reason and taste are the grand prerogatives of European minds and 'though we cannot agree with the sage preceptor (Aristotle) of that ambitious Prince (Alexander) that the Asiaticks are born to be slaves yet the Athenian poet seems perfectly in the right when he represents Europe as a sovereign princess and Asia as her handmaid.'[32] Jones also had contempt for Indian sciences: Jones remarked in the second annual discourse : 'Permit me now to add a few words on the sciences properly so named: in which it must be admitted that the Asiatics if compared with our western nations are mere children.' He also dismissed as 'almost entirely lost unfortunately for the curious European but happily for the patient Hindu since a revealed science precludes improvement from experience to which that of medicine ought above all others, be left perpetually open. For the nature of western appropriation of Ayurveda in colonial Bengal and its real status in the early period, reference may be made to my paper on the subject.[33]

He was introduced to Sakuntala by Pandit Radhakanta who told him that Natakas were like English plays performed in Calcutta. He read the Bengali recension of the drama with the help of Pandit Ramlochan and made a Latin and then an English translation from this brief acquaintance with the text. He hailed Kalidasa as the Shakespeare of India, a wild comparison in terms of space and time but we are still elated. This was a European discovery in their own terms for their own consumption.

Oriental Studies : The Future

William Jones was at the crossroads of oriental studies and orientalism. Whatever he did and for whatever purpose and however imperfectly, he introduced Indian lore to the West and earned respectibility for India in western eyes. Indians suffering from total self-abnegation under colonial rule learnt self-respect. The Indian discovery of India began ever since Jones went to the grave. Raja Rammohan Ray translated five Upanishads, wrote the Tuhfat and composed the Gaudiya Vyakaran almost unnoticed. Akshay Kumar Datta wrote the History of the Indian Saints which is beginning to be appreciated now. Raja Radhakanta Dev compiled Sabdakalpa drama. It is true that Indian Orientalists tool their cue from *Oriental Jones* to launch their indigenous Indology as late as 1885 when Raja Rajendralal Mitra became the first Indian President of the Asiatic Society heralding the Indianisation of the Society. He was followed by Haraprasad Sastri, Rakhaldas Bandyopadhyay and others. A rewriting of Oriental Studies must begin with republication and assessment of the works of the Pandits of the 18th century. Works of Rammohan, Akshay Datta and Radhakanta Dev, Ramkamal Sen, Krishnamohan Bandyopadhyay and even Tarachand Chakrabarty have to be reconsidered as part of Indian Oriental Studies. Indian Indology since Rajendralal have to be re-evaluated even by comparing them with the rudimentary efforts of a band of European Orientalists. The Asiatic Society as an institution of national importance can undertake this as its future project by being a little less worshipful of its founding fathers.

References

1. David Kopf, British Orientalism and the Bengal Renaissance, Calcutta 1969, p. 275.
2. Michel Foucault, The Order of Things, London, 1970, p. 372.
3. Michel Foucault, The Archaeology of Knowledge, London 1972, pp. 64-68, 183.
4. Antonio Gramsci, Prison Notebooks, tr. by Hoare and Smith, London, 1971, Introduction.
5. E.H. Said, Orientalism, London, 1978 p. 12.
6. *Ibid.*, p. 78.
7. Bryan.S.Turner, Marx and the End of Orientalism, London, 1978 passim.
8. S.N. Mukherjee, William Jones, A study in Eighteenth Century British Attitudes to India, Bombay, 1987 Second Edition, Introduction p. 3.
9. B.S. Cohn, 'The Command of Language and the Language of Command' in Subaltern Studies IV, edited by Ranajit Guha, Delhi, 1985, p. 316.
10. Ronald Inden, Imagining India, Oxford, 1990; Introduction.
11. Garland Cannon, The Life and Mind of Oriental Jones, Cambridge, 1990, p. 185
12. Garland Cannon, Letters of Sir William Jones, Oxford, 1970, Vol. 2 p. 601.
13. Cannon, The Life etc. p. 184.
14. *Ibid.*, pp. 184-192.
15. B.S. Cohn, infra p. 315.
16. S.N. Mukherjee *op.cit.*, p. 69.

17. Quoted in *Ibid.*, p. 74.

18. *Ibid.*, p. 73.

19. *Ibid.*

20. O.P. Kejariwal, The Asiatic Society of Bengal and the Discovery of India's Past 1784-1838, Delhi, 1988, p. 40.

21. *Ibid.*, p. 32.

22. *Ibid.*, pp. 226-27 Cf. Abhijit Mukherjee, Science, Empire and the Asiatic Society, 1784-1855 in The Historical Review Vol. I (New Series) Nos. 1 & 2 1992, forthcoming in C. Palit ed. Revolt Studies 3.

23. Abhijit Mukherjee, 'European Jones and Asiatic Pandits' in Journal of the Asiatic Society Vol. XXVII No. 1, 1985 passim.

24. Cannon, Oriental Jones, Boston, 1964 p. 131.

25. Rosane Rocher, Orientalism, Poetry and the Millennium: The Checkered Life of Nathaniel Brassey Halhed 1751-1830, Delhi, 1983 p.65 f.n.

26. A. Mukherjee *op.cit.*, p. 49 quoting Asiatic Researches Vol. I p. 299.

27. A.R. Vol. II pp. 111-47.

28. A. Mukherjee *op.cit.*, p. 53.

29. *Ibid.*, p. 54 quoting A.R. Vol. II pp. 225-28.

30. *Ibid.*

31. A.R. Vol. I, pp. 422-23.

32. A.R. Vol. I 2nd Discourse quoted in S.N. Mukherjee *op.cit.*, p.109.

33. C. Palit, 'Ayurveda in Colonial Bengal' (forthcoming in Scientific Bengal).

3

Serampore Missionaries and Science

Introduction

With the onset of British rule in the late 18th century, India was opened for missionary work on a large scale. Missionaries believed that the Indians were eager for precepts of Jesus and consequent conversion to Christianity though there was no basis for it. The Baptist missionaries since their inception in 1782 nurtured these ideas. Many marginal people in England looked to the Empire for occupation and turned to missionary work in India. Conversion would also eventually create a group of people loyal to the Empire.

The second motivation was to spread the Christian Civilization in the colony as white man's burden. The colonial culture was held barbarous and Christian Civilization superior without any rationale, particularly when India had an ancient civilization. Its purpose was to culturally convert the Indians and make them grateful to the colonial masters Charles Grant, a director of the East India Company wrote a pamphlet to this effect and thus became a prophet of evangelism in 1792. The real prophet of evangelism was William Wilberforce who after the Charter Act of 1813 wanted to utilise India as a field of missions for both religious and cultural purposes.

Missionaries serve as an auxiliary force for collecting all kinds of information for the Empire. It helped consolidation of the administration and colonial extraction of primary resources. The missionaries were therefore inextricably linked with colonialism but in the process, consciously or unconsciously, they also served the Indian people and left a valuable legacy of science and culture.

Serampore Missionaries William Carey

The work of Serampore missionaries has to be reviewed in this background. One of the founding fathers of the Serampore Mission was Willam Carey. Carey came from a very humble origin in Northumbria. His father was a weaver, later a village school teacher, and then a clerk in the local church. Carey was a very attentive student and keen observer of surrounding nature. At an early age, he picked up Latin and started learning Greek and Hebrew.[1] But due to poverty, he gave up studies and took up the job of a cobbler. He had a friend in John

Ware who inducted him to missionary work.[2] He joined the Baptist Society in 1782 and became a priest in 1789. From 1791 he wanted to go abroad for missionary work and in 1792 he joined the Baptist Mission.[3] He met Rev. John Thomas who had returned from Bengal and decided to come to Bengal. It took some time to collect expenses and permission of the East India Company. He reached Calcutta on 11 November, 1794 and took Ramram Basu as his assistant. But he got no job in Calcutta and finally landed in Malda to work in the indigo factory of Woodney at Madnavati.[4] But this factory was closed down. At this time, Serampore Baptist Mission was also established on 25 December 1799. William Carey got a call from William Ward to join the Mission.[5] He also got an offer from Fort William College to teach Bengali and Sanskrit.

William Carey was a versatile scholar. He got the Bible translated in thirtytwo regional languages. He also got Ramayana translated for the first time and published the gist of the Vedas. It is not the purpose here to review his literary work. We turn to his scientific work and achievement. While at Madnavati, Carey raised a small botanical garden for botanical research. His main attempt was to grow European fruits on Indian soil. He was successful in this experiment. At Serampore he set up a larger botanical garden on 5 acres of land. Botanists considered this garden superior to Sibpur BotanicalGardens. About his garden, Prof. Bruhl commented that most of the plants in Bengal had grown out of the seeds of the plants in Carey's garden carried by birds. There were 427 types of plants in Carey's garden arranged in the Linnean system.[6] It was a full-blown garden of numerous fruits and flowers, groves and aviaries and tanks making it a natural museum.[7] It had a central path known as Carey's walk. This has found mention in Voygt's Hortus Suburbanus Calcattensis compiled from notes in Carey's diaries. Carey had also contributed to the plant family in the Sibpur Botanical Gardens. He was in intimate touch with his friend Roxburgh, Superintendent of the Gardens. Roxburgh named a Sal tree Careya Saulea.[8] But Carey changed its name to Shorea Robusta. But Roxburgh in gratitude to his friend named two groves of medicinal plants Careya Arborea and Careya Sphoerica. Roxburgh prepared a catalogue of 3,500 trees in Bengal. Carey edited this volume in 1814 and named it Hortus Bengalensis. Carey wrote about the improvements that would be made in the Sibpur Botanical gardens. Roxburgh left another monumental work, Flora Indica with Carey before leaving India, Carey also edited this work together with the next Superintendent of the Gardens, Wallich. This was published in three volumes between 1820 and 1824. A second edition of this work came out in 1832. Carey also edited Wallich's Asiatic Plants. Thus Carey can claim to be the father of botanical studies in India.

Apart from Botany, Carey had first experience in agriculture. On arrival in India, he attempted cultivation in Sundarbans and also in Malda. He was a keen observer of agriculture in Bengal and in 1812 his famous essay. 'The State of Agriculture in the District of

Dinajpur' was published in Asiatic Researches. In it, he pointed out the low level of agriculture in Bengal and the ways to improve it. With the latter purpose in view, he founded the Agri-horticultural Society of Eastern India on 14 September 1820. The Governor General acted as a patron. In its agenda, he suggested, among other things, compilation of all kinds of information about Indian agriculture, improvement of agriculture by the use of better manure, rotation of crops, irrigation, drainage and scientific agricultural implements, improvements of cattle and bringing uncultivated land under cultivation. He and Ramkamal Sen were the first joint secretaries of the Society. In 1824 he became the president of the society and in his presidential address, he pointed out that India's agriculture was behind England by 200 years. The Government must take part in promoting agriculture in the districts. With that object in view, questionnaires were sent to the experts in the districts. Radhakanta Dev answered the questions for 24 parganas. In this way, huge information was collected about Bengal agriculture. The society used to answer the queries of cultivators and supply seeds, plants and manures to them. Agricultural fairs were held and cultivators were awarded prizes for their performances. The ideas spread to other cities in India and even to Singapore. Carey made it an agricultural movement. A journal was published for agriculture first in English and then in Bengali. Carey was also instrumental in the establishment of plantation committee and was its first secretary. Its main work was afforestration for production of timber. Thus Carey's contribution to Indian agriculture was immense.[9]

To popularise science in Serampore College, Carey evolved a new education system in 1813. Previously religion, linguistics, history and geography were taught here. Now in the Serampore system, many scientific subjects were introduced and they were taught in Bengali. Physics, Meteorology, Chemistry, Botany, Geology and Biology were introduced. Small scientific works were written for the purpose and charts and questionnaires were used. In collaboration with Calcutta School Book Society, many scientific books were published. Of them, Felix Carey's Physiology, Marshman's Astronomy, Mack's Chemistry are well known. Besides, newspapers and periodicals like Samachar Darpan and Digdarshan carried many scientific articles. The Baptist Mission established over thousand schools in North India catering to ten thousand students reading in both Bengali and Hindi.[10]

Serampore College was established in 1818. Here literature of the East and sciences of the West were taught. Teaching was backed up by well-stocked library, laboratory, museum, medical unit and observatory. Carey taught botany, zoology and agriculture. The famous scientist John Mack taught chemistry from 1821 onwards.[11]

Technology was not neglected, Carey had constructed a wooden press at Madnavati. He was helped by Panchanan and Manohar Karmakar. The trio made a type foundry here. From this press, two lakh books were printed during Carey's lifetime. For this purpose, a paper mill was set up in 1820. Good quality paper was produced here from local material. To start with, work was done manually. Later a rotary machine was imported from England at a cost of Rs. 30,000/-. Thus local people were trained in the art of paper manufacture. This paper was cheap and durable. After Carey's death in 1826, John Marshman continued the mill but in 1845 Government stopped it by an act instigated by the British manufacturers of paper. By then Serampore had considerably contributed to paper and printing technology in India.[12]

Carey's contribution to science and technology in India is immense. But it did not explore Indian tradition of science. He organised Indian botany in western terms. The main purpose was to find out valuable cash crops like good timber, medicinal plants, cotton, silk, jute and grow them for export. Cereals, fruits, flowers, vegetables were also experimented with foreign seeds. The Agri-horticultural society founded by him tried to improve agriculture by this means. Carey was critical of the standard of agriculture in this country and blamed the local peasants and zamindars. But he never criticized colonial revenue policy which impoverished people and depressed agriculture.

Admittedly Carey caused a printing revolution by printing two lakh books. But the major part of it was translation of the Bible in 40 languages used for missionary propaganda. But the books on literature and the journals contributed to Bengal renaissance. The translation works of Carey were in many cases carried out by Bengali scholars like Ramram Basu, Mrityunjoy Vidyalankar and Rajiblochan Mukhopadhyay. Similarly the contribution of Panchanan and Manohar Karmakar in the field of printing technology has to be duly acknowledged.

The scientific works issued by the Serampore Mission were not always scientific. Many scientific explanations of creation were often brushed aside and explained in terms of God's grace. Translations were not always correct and to the point and the language was faulty. Yet, Carey was undoubtedly one of the heralds of Bengal renaissance and science popularisation in Bengal.

Felix Carey (1776-1822)

Felix Carey was William Carey's son. He came to India with his father. Despite his best wishes, William could not provide good education for Felix. He also lost his mother early in his life. Felix learnt Bengali from his father and on his own picked up Sanskrit and Hindustani. At Serampore, Felix assisted Rev. Ward in the work of the press. He was also connected with missionary work and running of schools. Later he was attracted to medicine under the influence of Dr. Taylor, a missionary doctor at Serampore. Apart from his apprenticeship with Dr. Taylor, he took practical training in medicine in a Calcutta hospital.[13] After a period of time, he went on missionary work to Burma. Here he learnt Pali language and began to write a dictionary in English for that language. He left his job in Burma after 3 years and returned to Serampore. During this time, he wrote a few valuable scientific works in Bengali.[14] He died at the raw age of 36 in 1822. The scientific work for which he had attained abiding fame was Vidyaharabali. This was a kind of encyclopaedia. Felix had planned to discuss one subject in each volume. The first volume was called anatomy. This is the first Bengali book on physiology. It was completed in 14 chapters. He also translated John Mack's Principles of Chemistry in Bengali and wrote a few essays on scientific subjects.[15] Sajanikanta Das has hailed him as one of the major foreign writers of Bengali language and literature.

John Mack (1797-1845)

John Mack was born at Edinburgh. He was a very brilliant student at school and college. He took a master's degree in science from Edinburgh University. He was gifted in mathematics and chemistry and proficient in Greek and Latin. He had a command over all branches of science and knew the latest information about them. He then joined the Church of Scotland to become a missionary and then left it in favour of the Baptist Mission in Bristol. Rev. Ward invited him to Serampore and he joined Serampore Missionary College as Professor of Science.[16] He carried a lot of scientific books and apparatus with him from Scotland which he got from his friend Douglas. He set up a laboratory with these at Serampore. This was the first good scientific laboratory in India. Though he carried out his research in chemistry primarily, his main interest was to popularise western scientific theories among Indian students. He used to teach by scientific experiments with the help of his apparatus. This type of teaching was introduced in India for the first time and attracted students from different parts to Serampore. Serampore College became an advanced centre of learning and Mack's fame spread far and wide. Lord Hastings invited him to give a course of lectures on Chemistry at the Asiatic Society. After these lectures, he was allotted a room for his use at the Society and got an award of a £105 which he donated to the Mission.[17]

He got interested in writing scientific books in Bengali together with Marshman. His Principles of Chemistry also came out in Bengali. This was the first Bengali book on chemistry. He created suitable scientific terminology in Bengali for the understanding of the local people. At that time, Indian scholars could not dream of writing a scientific work in Bengali. He also prepared a map of India in both English and Bengali and gave it to Lord Hastings. This was later lost. Besides Mack was assistant to Marshman for the publiction of Serampore newspapers. On Carey's death and after Marshman became ill, Mack became the principal of Serampore College in 1819. He also had the charge of the Mission in addition. He died on 30 April, 1845. Mack's scientific work was revolutionary. Though his Bengali later became oldish, this was followed as a text by the Bengali readers.

Conclusion

The role played by the Serampore Missionaries in the spread of science in India was undoubtedly path-breaking. But its colonial context should not also be forgotten. Along with religious conversion, cultural conversion went hand in hand. Its purpose was to ingratiate Indians with British rule. To prepare India for colonial administration and extraction of natural resources, some amount of scientisation was necessary. Though Serampore missionaries were not directly linked with it, their work indirectly helped British colonialism. But that should not take away from the good deeds done by the missionaries in the field of science. Indians definitely benefited from their efforts.

References

1. Drewery Mary, William Carry. Shoelmaker and Missionary, Kent, 1978, p. 25.
2. Drewery, *op.cit.*, pp. 15-19.
3. Drewery, *op.cit.*, pp. 39-42.
4. 'William Carey's Journal', June 11-June 18, 1794.
5. Drewery, *op.cit.*, pp. 109-110.
6. Ghosh, A.K. and Datta R., 'History of teaching Botany in Bengal', The Journal of the Asiatic Society of Bengal XVIII, 1976, p. 8.

7. Ghosh A.K., 'William Carey', in Science and Culture, Dec. 1994, p. 222.

8. Chatterjee, Sunil Kumar, Contribution of Dr. William Carey towards the Development of Indian Science and Technology, prog. BITM Seminar, on Calcutta and science 1978.

9. Ghosh A.K., *op.cit.*, p. 223, Serampore Mission, First Report of Native Schools 1818, p. 51, S.K. Chatterjee, William Carey and Serampore, Calcutta 1984, p. 14.

10. Serampore Mission, Third Report of Native Education 1820.

11. Banerjee B.N. Sambad Patrey Sakaler Katha (in Bengali) Calcutta, 1377 BS. pp. 19-20.

12. Chatterjee S.K., William Carey and Serampore, *op.cit.*, p. 41, also Serampore Missionaries, Baptist Missionary Society Periodical Accounts, Vol. IV, 1811-12; Khan, M. S., Bangla Mudran O Prakasany Carey (in Bengali) Dhaka, 1964, p. 76; Serampore College Annual Report 1824, E.D. Potts, Serampore Press and its printers (unpublished).

13. Chatterjee, Sumil Kumar, Felix Carey (A Tiger Tamed), Serampore, 1991, p. 24.

14. *Ibid.*, pp. 35-47.

15. *Ibid.*, p. 82; Das, Sajani Kanta, Bangla Gadya Sahityer Itihas in Sahitya Sadhak Charitamata Vol. 88, p. 32.

16. Chatterjee, S.K., Rev. John Mack, The first professor of Science in modern India, Calcutta, 1998 pp. 21-22.

17. *Ibid.*, pp. 23-25.

4

Science and Nationalism in Bengal (1876-1912)

The nationalist stand on science in Bengal can only be understood if we correctly define the current expression in this area, colonial science. Most academics understand by it only the time-frame, i.e., science that developed during the colonial period. Many include even the nationalist science activities and achievements as a product of colonial rule. George Basalla goes further to mean that colonial science is the arrival and acceptance of European science in the colonies, where there was not a semblance of science before European colonisation of India. It led to transfer of science and technology for the first time and its dissemination in the benighted country and its consequence was her material progress.[1] Both are unhappy explanations of colonial science and Basalla's Eurocentric view can not go unchallenged. It is common knowledge that ancient civilizations like

India had a rich scientific past before the British conquered her and the tradition was not wholly lost though it was hegemonised by Imperial Science. What happened in British India was not transfer of science and technology but transplantation of European science and technology. Colonial science was not, therefore, a matter of time-frame or periodisation of history. It meant much more than that. Colonial science in my opinion is accumulation of science knowledge about the resources of the colony and utilisation of this knowledge for extraction of these resources for colonial purpose. In this sense, the Baconian dictum that 'knowledge is power' can be recorded.[2] Foucault more incisively makes this equation of knowledge with power and its control and manipulation for the benefit of the ruling class.[3]

Right from the beginning, the British authorities had no intention of imparting scientific education to Indians. The pursuit of science was confined to the Asiatic Society and the various survey offices. These were closed preserves of the Europeans and Indians had only a nominal participation as subalterns.[4] The Asiatic Society was the citadel of colonial science. It gathered all kinds of scientific information about India by painstaking researches done by the Company's British officials, based on information supplied by the Indian munshis and the pundits. This information was doubtlessly fed into the administration for colonial extraction. The survey officers served the same purpose. The land survey, botanical survey, geological survey, anthropological survey offices had qualified personnel in science but the information that they gathered was closely guarded for colonial use.[5] The knowledge was not for dissemination to the Indian public. This is the reason why at the time of Anglicist-Orientalist controversy : Rammohun wanted the sum of one lakh rupees allocated by the Company's Government for imparting scientific education to the Indians. In his letter to Lord Amherst in 1823, he clearly pointed out that India was already rich in philosophy and literature but deficient in the scientific knowledge of the West. He implored the Government to spend the sum on promotion of natural sciences in educational institutions.[6] For this purpose scientific books had to be imported, laboratories had to be set up and qualified foreign teachers from abroad had to be appointed. But this appeal of Rammohun had fallen on deaf ears and ultimately in 1835, Macaulay settled the contest by having an Educational Minute passed in the council in favour of English education but mostly literary. In the Government colleges and schools the literary curriculum prevailed. Whatever science was taught in the Hindu College was of an elementary nature.[7] Shivnath Shastri in his memories narrates how one Professor Tytler used to teach only the composition of soda in his Chemistry class and thereby was nicknamed 'Soda Sir'.[8] The Calcutta Medical College was the only academic institution where science was taught at a high level because medical students would have to deal with life and death questions of even white civilians.[9]

Dr. Mahendra Lal Sircar was born in 1833, the year of Rammohun's death and later destined to become the father of modern science in India was admitted to the Hindu College initially. But determined to pursue a science course, he left the Hindu College to join Calcutta Medical College for this purpose much to the surprise of his guardians and classmates who still thought that a literary English education was not only a bread-winner but also the most civilized form of education.[10]

This brings us to a study of the phenomenal career of Mahendralal who rose to become not only the top homoeopath of his time but also the father of Indian Science. As we have already observed, he joined the Medical College where he could study science proper to his intense delight. He duly qualified for L.M.S. and in 1863 he was the second Indian M.D. from the College.[11] He set up his medical practice and became an active member of the Calcutta Chapter of the British Medical Association. In his initial years, he was a doughty champion of Allopathy and in one of the meetings of the association he denounced Hemoeopathy as mere quackery.[12] At this stage he got Morgan's Philosophy of Homoeopathy for review in the *Indian Field*. On reading this work he got converted to Homoeopathy and wanted to know more about it. He became an apprentice with Dr. Rajendralal Datta, the legendary Homoeopath of Calcutta. The principles and practice of Homoeopathy convinced him of the efficacy of Homoeopathy as a cure.[13] He placed it above Allopathy in a seminar of the British Medical Association. It raised a hue and cry among the charmed circle of the leading Allopaths of the city. Ewert, Waller and Surya Kumar Goodeve Chakravarty. They launched a tirade against him for his apostasy. He was declared to be a mad man and he was ostracised by the doctor community. Mahendralal began to lose his practice and was put to dire financial straits. Even his non-paying clients deserted him. But Mohendralal persevered. Those who had left him in favour of allopathy began to return slowly. His practice looked up again. The dark period was over. He made such a name in Homoeopathy that his fame was little short of Dr. Rajendralal Datta.[15]

Mahendralal had not forgotten the conspiracy of the British doctors to ostracise him for his switching over to Homoeopathy. This had injured his freedom to pursue Homoeopathy as an alternative science. He considered it as a colonial constraint and since then contemplated the founding of a national science association which would be controlled and funded exclusively by the Indians themselves. The projected association would be dedicated to the cause of science in India for the purpose of national reconstruction.[16] Mahendralal started his campaign for the association from 1867 onwards. He published appeals in the Hindu Patriot, Calcutta Medical Journal etc. and issued pamphlets during his public lectures on this subject. Public opinion in favour of the project was fast building up.[17] But there was a crunch of funds. Mahendralal sent out appeals to Rajas, Maharajas, Zamindars and professional elites of India to contribute liberally to this noble cause. The response was overwhelming. To this Mahendralal contributed his life's savings.[18]

Once this launching pad was ready, he appealed to the Bengal Government for support to the cause. But at this stage, he faced a stiff opposition from an unexpected quarter. The Indian League which had branched off from the British Indian Association as an organ of the professional landed middle class was seeking recognition from the Bengal Government. The Ghosh brothers, Motilal and Sisir who were the founders of the Indian League wanted to placate the British Government, particularly Lt. Governor Sir Richard Temple.[19] They came out with a rival proposal of establishing a technical institute to be named after Prince Albert, consort of Queen Vistoria who was then visiting India. The argument that they gave was that an institution for vocational education was more desirable to provide employment to poor and jobless people. A science association could wait as it was the territory of the privileged. The colonial Government would only be delighted if the national science association was thwarted by inner conflict of this sort. The Indian League Govenment alliance came to be popularly known as the Ghosh Temple League.[20] Both sides began to mobilise their support from the public. Ultimately, Temple sought to resolve these conflicts by holding a public meeting in the Town Hall. The historic meeting was held in 1876. Rev. Eugene Lafont of the St. Xaviers College, Keshab Chandra Sen and Rajendra Lal Mitra took the side of Dr. Mahendralal Sircar in this debate while Sambhu

Chandra Mukherjee, the noted editor of Reis and Ryot and veteran Rev. Krishna Mohan Banerjee took the other side. The eloquence of Keshab Chandra and the arguments of Father Lafont and Rajendra Lal Mitra won the day for the National Science Association.[21] Lafont pointed out that the National Science Association was not exactly opposed to the programme of the technical institute. India needed a pool of scientists who would also act as trainers of the artisans at the technical institute. Otherwise they had to depend on foreign experts. Rajendra Lal Mitra supported Lafont on this score by saying that only a generation of Indian scientists could train our artisans to higher skill and innovation.[22] Mahendralal himself replied effectively by saying that India could produce able scientists if not Galileo, Newton. Herschel and the like. The House voted in favour of the association and Temple had to accept the verdict. It was decided by the Government that a house at 210, Bowbazar street would be alloted to the National Science Association on the security deposit of Rs. 50,000 out of the total fund of Rs. 70,000 to be raised by subscription by the association. Thus the historic Indian Association for the Cultivation of Science (IACS) was launched in 1876.[23]

Review of IACS : (Indian Association for the Cultivation of Science). The IACS was dedicated to the promotion of higher studies in science and popular science. Mahendralal had hoped that the IACS would one day become an advanced centre of scientific research like the Royal Institute in England. The IACS, therefore, was planned accordingly. Its purpose was to absorb the benefits of western sciences and be abreast with it. But he was also for the utilisation of the saving elements of traditional science in India. Both were necessary for national reconstruction. Mahendralal also arranged a series of popular lectures on various scientific subjects which were generally held in the evening with the help of lantern slides. The purpose of these lectures was to popularise science among the common people. They were a big draw for college students who used to throng the hall in large numbers. The IACS was nicknamed a private college of science for this reason. But its efficacy cannot be gainsaid.[24] It led to considerable popularisation of science and Mahendralal's hope to fight superstition with science was largely fulfilled. The lecture domonstrations also had a practical side and promoted vocational training. The IACS had a long record of involving all noted scientists of Bengal starting from Mahendralal himself, Father Lafont, J.C. Bose, P.C. Ray, Nilratan Sarkar, Asuthosh Mukherjee et al., who contributed to both higher and popular science. The objectives of creating a pool of Indian scientists was also achieved. Mahendralal in his deathbed testament had lamented that the IACS could not become like the Royal Institute because of lack of patronage from the Indian aristocracy. But this frustration has no basis. After his death, the IACS expanded its activities and went from strength to strength. During the era of C.V. Raman, Meghnad Saha and K.S. Krishnan in the 30s of this century, its prestige was no less than that of the Royal

Institute. All three contributed significantly to higher science and Sir C.V. Raman went on to become the first Indian Nobel Laureate in science.[25] Now it has attained world wide recognition as an advanced centre of scientific research and a colony of well-established Indian scientists, who can train all apprentices in the principles of modern science and technology. Indian dependence on foreign scientists has virtually been eliminated in the field of industrial recovery and reconstruction. Our society has been considerably scientised, though its contribution is still restricted to the urban sector.

The Dawn Society

The lead given by the IACS in the field of nationalist pursuit of science was carried forward by the Dawn Society founded by Satish Chandra Mukherjee in 1902. Satish Chandra was already famous as an educationist. A teacher of South Suburban School in Bhowanipore, he had already established the Bhagavat Chatushpathi in 1895 which undertook studies on religion and philosophy. Its purpose was to explore the cultural heritage of India. In 1897, he launched the Dawn Magazine in which he propagated his ideas of national education. It was Satish Chandra's firm conviction that to love one's country one had to know it thoroughly. He started a column entitled 'Indiana' in the Dawn Magazine carrying numerous articles on different aspects of Indian civilization. The Dawn Magazine was widely circulated in every educated household as a journal of high standard and taste. In its run between 1902-1906, the Dawn took up the cause of national education in view of the official assault on higher education. The Government sought to control the University Senate and Syndicate through official nomination and shrinking educational opportunities by disaffiliation of indigenous colleges. Satish Chandra criticised the syllabus of Calcutta University as "all-too-academic, all-too-literary, unscientific and unindustrial".[26] In 1902 the Dawn society was established in the premises of the Metropolitan College with a new teaching syllabus and schedule. The courses were divided between general and industrial/commercial. Most classes were like practical classes where students were encouraged to participate in group discussion and submit their notes taken from lectures for scrutiny by teachers.[27]

The Dawn Society had a definite syllabus in science and technology. It had the four-fold objectives of salvaging the worthy elements of traditional science, absorbing the benefits of modern science, spreading science to masses and offering job-oriented technical education. In this sense it continued the tradition started by Mahendralal at the IACS. Along with teaching science and technology from lower to higher levels, the members of the Dawn society wrote profusely on scientific subjects in the columns of the Dawn magazine. From 1896 no less than 70 essays appeared in the Dawn on science, technology and allied subjects. Members of the Dawn Society comprising of eminent scientists of the day wrote these articles.[28] Satish Chandra himself was responsible for many editorials on scientific subjects.

The other important sector of the society's activities was the Industrial School run by it. It was like a technical institute where training was given by experts for various kinds of technical jobs of the elementary level. This was a programme which Binoy Sarkar later called 'Mistrification'.[29] In connection with the Bengal Technical Institute, the Dawn Society produced many fitters, jointers, welders, chemists and weavers at the Industrial school. The products of these trainees were also displayed and sold in exhibition and in shops in Burrabazar. The sale figures show that the industrial school performed well. It opened up job opportunities for the new generation. Thus the national science movement not only continued in the proceedings of the Dawn Society but it also served as a vital link between the IACS and the future National Council of Education wich founded the Bengal Technical Institute and finally the College of Engineering and Technology. To that historical development, we turn next.[30]

The NCE (National Council of Education)

Bengal was partitioned in 1905 to divide and weaken the Bengali Bhadraloks who were taking leading part in the freedom movement in Bengal. It caused a consternation among the educated Bengalis and they decided on a political movement to annul the partition at all costs. The movement comprised of boycott, swadeshi and national education. The programme involved national education for swadeshi industrialisation. It meant a curriculum of science and technology for national reconstruction. The Dawn Society had already started such a course. It got merged in the activities of the National Council of Education which was founded in 1906 by the nationalists for national education including science and technology. When Satish Chandra founded the Dawn Society, its purpose was to spread national education when the Government of Bengal was trying to shrink the opportunities created by educated Bengalis. The National Council of Education had to face the further problem of government intervention because of its swadeshi activities.[31] The government banned all kinds of nationalist activity including singing of national and patriotic songs in schools run by the Council. Faced with this kind of oppression, the NEC got determined to purpose its ideals of national education more vigorously.

Its programme of boycott and swadeshi led it to provide an alternative path of scientific and technical education for swadeshi industrialisation. For this purpose, along with the Bengal National College which taught general arts and science, it established the Bengal Technical Institute (BTI) which provided the first phase of technical education.[32] This has been described as the programme of bread and butter education or 'Mistrification' or the first phase of industrialisation by Binoy Kumar Sarkar. The fathers of NCE emphasised on nationalism as the main incentive for the alternative form of education. But those who joined the BTI would not mix politics with their education. Sir Tarak Nath Palit who had donated 10 lakhs and his residential house at 92, Upper Circular Road also thought that the colleges under NCE should not be branded as political hotbeds.[33]

As soon as the Bengal partition was annulled in 1911, the nationalist euphoria also abated. Sir T. N. Palit withdrew his munificence and gave it to the Calcutta University which had Sir Asutosh Mukherjee as the new Vice-Chancellor appointed by the Government. Sir Asutosh did not abandon the nationalist path, but Calcutta University was not known to be a centre of political agitation. At least the Bengal Government thought it to be a safe place and Sir Asutosh a sage steward. Sir Asutosh absorbed much of the programme of the national education of the NCE without giving offence to the Government.[34] The net result of this overtures was that the Bengal National College and the BTI come under fire. The BTI had to change its venue but it gained popularity amongst students who did not get any opportunity to have technical education under Calcutta University. By 1917 it began to draw more students that the Bengal National College which looked deserted since 1911. The BTI had a successful run down to 1928 and provided valuable vocational education to many who got settled in life. From after 1928, the NCE transformed it into the College of Engineering and Technology (CET) which became the second phase of technological and engineering education or in the words of Binoy Sarkar, "the phase of industrialisation in education".[35] It became a close rival of the B.E. College,

Shibpur for its variety of engineering subjects offered like Mechanical Engineering and Chemical Engineering. From then on it never looked back. It got illustrious teachers in place of those who had left for Calcutta University. They created a nationalist tradition from the generation of Hiralal Roy to Triguna Sen. Its alumni went on to become the architects of modern India. Many went abroad and settled there having name and fame. Many established swadeshi industries like the Bengal Lamp, Bengal Waterproof etc. Binoy Sarkar calls this phase "Modernisation".[36] Thus nationalism and science went hand in hand in creating a swadeshi base of industrialisation and a pool of indigenous scientists to run it. Calcutta University under Sir Asutosh in its Colleges of Science and technology also performed well. But it still suffers from the trapping of colonial education whereas National Council of Education was imbued with a nationalist spirit till the time of its renaming as Jadavpur University in 1956. The nationalist spirit ebbed obviously after independence but it had already played its glorious innings in swadeshi reconstruction under the withering touch of colonial rule.

References

1. George Basalla : 'The Spread of Western Science'. *Science*, 156, 1967. pp. 612-13.
2. E. W. Said : Orientalism, London, 1978, p. 12.
3. Michel Foucault : The Order of Things, London, 1970, 372: also, The Archaeology of Knowledge, London, 1972, chapter VI, 64-68 and 183.

4. C. Palit, William Jones, Orientalism and Science, in JASB, 1996, Calcutta.

5. Deepak Kumar : Science and the Raj, Delhi, 1996.

6. A.F. Salauddin Ahmed : Social Ideas and Social changes in Bengal, 1818-35, Calcutta, 1976, p. 159.

7. H. Sharp : Selections from Educational Records, 1781-1839 Part-I, Calcutta, 1920, p. 317.

8. Shivnath Sastri : Ramtanu Lahiri O Tatkalin Banga Samaj (in Bengali) passim.

9. A.C. Ukil : Centenary of the Calcutta Medical College, 1835-1934, Calcutta, passim.

10. S.C. Ghosh : Life of Dr. Mahendralal Sircar, Calcutta, 1909, pp. 2, 5.

11. *Ibid.*

12. Dr. M. L. Sircar : On the supposed uncertainity in Medical Science and on the Relation between Diseases and their Remedial Agents, Calcutta, 1903, preface p. VI-VII.

13. *Ibid.*, p. viii.

14. *Ibid.*, pp. x-xi.

15. *Ibid.*, p. xvii.

16. The Englishman, 29 December, 1869.

17. Hindu Patriot, 13th December, 1869.

18. Dr. M.L. Sircar, Indian Association for the cultivation of Sciences, Calcutta 1877, preface p. VII.

19. *Ibid.*

20. Samarendra Nath Sen : Vijnanarchaya Dr. Mahendralal Sircar. (in Bengali), Calcutta, 1985. p. 49.

21. *Ibid.*

22. *Ibid.*
23. Arun Kumar Biswas : Science in India, Calcutta 1969, p. 55.
24. A Century (Centenary Papers of The Indian Association for the Cultivation of Science), Calcutta, 1976. Chapter VI, see the Records.
25. H & U Mukherjee : The Origins of the National Education Movement, Calcutta 1957 p. 7 also The Dawn, February, 1898 p. 354.
26. *Ibid.* p. 319-322.
27. C. Palit : A Selection of Scientific Papers of the Dawn Society 1897-1912 (forthcoming).
28. Binoy Kumar Sarkar : Education for Industrialisation, Calcutta, 1946, 99, 94-97.
29. *Ibid.* passim.
30. Minute Book 1906-1924 of National Council of Education 3rd January, 1907.
31. Binoy Kumar Sarkar *op.cit.*, pp. 86-94.
32. Annual report of national Council of Education, 1912, p. 4.
33. Binoy Sarkar *op.cit.*
34. *Ibid.*, p. 18.
35. *Ibid.*
36. N.K. Sinha : Sir Asutosh Mukherjee, Calcutta, passim.s

SECTION-II : TECHNOLOGY

5

Sitanath Ghosh-The Forgotten Pioneer of Technology in Bengal

The Setting

The Manifesto of the Hindu Mela founded on 12th April 1867 can be traced to the 'Prospectus of a Society for the Promotion of National Feeling among the Educated Natives of Bengal' written in 1866 by Rajnarain Bose, the doyen of Hindu Nationalism. Bose himself claims this in his autobiography. The Tagore family was behind the Mela from the beginning. It was Devendranath Tagore who launched the National Paper on 7th August 1866 and made Nabagopal Mitra its editor. The National Paper caried the prospectus in full and it occured to Nabagopal that its contents could be implemented through a national fair or what came to be known as Hindu Mela. Its first secretary was

Ganendranath Tagore, and Nabagopal was the assistant Secretary. Even Dwijendranath Tagore, the eldest son of Devendranath was one of the prime movers of the Mela. The prospectus observes: 'Now the European ideas have penetrated Bengal, the Bengali mind has been moved from the sleep of ages .... it is to be feared that the tide of revolution may sweep away whatever good we have inherited from our ancestors. To prevent the catastrophe and to give a national shape to reform, it is proposed that a society be established by the influential members of native society for the promotion of national feeling among the educated natives of Bengal.[1] Rajnarain mainly harps on Bengalee language and culture and does not say anything on swadeshi enterprise and science and technology. But from a speech of 1880, we gather that he had that issue very much in his mind: "Your arts and industries which won and are still winning the admiration of Europeans are languishing. Your artisans, the products of whose hands won such admiration are starving for want of employment. The immense material resources as well as the money of your country are being carried away by foreigners while we have to depend on England even for such a common article as lucifer match nay, even for the very salt which seasons our rice."[2] But Chandranath Bose said it more poignantly before the Social Science Association in 1869, 'She (India) is yet only a hewer of wood and drawer of water for English civilisation in the East. But once let manufacture be established in Bengal, let Bengal once know that the cloth which she wears, the paper on which she writes and the knife with which she cuts will be no longer prepared for her by England. Then will England herself form with her a friendship more close, more intellectual than subsists at present and then will that friendship be placed on that basis of mutual esteem and respect without which friendship is a serious misnomer.[3] The prophet of swadeshi, Bholanath Chandra wrote in Mookerjee's Magazines in 1873: The native English and Vernacular papers should preach for the founding of independent Native Banks, Native Companies and Corporations. Native Mills and Factories and Native Chambers of Commerce in the Presidencies. They should denounce the insensate practice of preferring foreign goods to home made manufactures.[4]

S & T in the Sessions of the Hindu Mela

The first session of the Mela was held in Duncan's villa in Belgachia on 12 April 1867. One of the six objectives was the exhibition of indigenous products by local artisans. The rest were on cultural lines as outlined in Bose's Prospectus. In the second session (1868), Ganandranath as its secretary stressed on self-reliance in all aspects of life and bewailed dependence on others. As announced in the inaugural sessions, exhibition of the swadeshi fine arts and handicrafts was organised along with a display of agricultural products and innovations. The effort was the first of its kind and therefore, the second session is so significant. Weaving and needle craft by women producing bags, shoes, mats and carpets, ornaments and ivory works by artisans and towels and dusters made by the prisoners of Alipore Jail were the main exhibits. The last mentioned were really durable and attractive products. Famous poet and playwright Monomohon Bose in his keynote address also underscored the importance of swadeshi goods made of indigeneous materials and skill uninhibited by the Europeans.

The third Session was held in the same villa in 1869 during Chaitra Sankranti as usual (hence also called Chaitra Mela). The industrial and agricultural exhibition included needle craft, woollen dresses, pottery, Benares silk, Chinese silk, gold and silver ornaments by goldsmiths of Dacca, musical instruments, weapons and statues. The main speaker, Monomohon Bose, however, pointed out that the needle craft etc. by women in imitation of the West would not serve the real purpose of self-reliance. Swadeshi needs and skill must be promoted first. The Mela adopted the practice of encouraging swadeshi art and craft by exhibition. But it provoked the Amrita Bazar Patrika to comment that the art and craft section was becoming the fancy fairs of the European ladies.

Appearence of Sitanath Ghosh

The fifth session (1871) was memorable from the point of view of technological innovation. Here in Hiralal Seal's garden, Sitanath Ghosh of Jessore demonstrated for the first time the cotton spindle he had invented by which yarn was produced and twisted mechanically in three places without manual effort at each turn though the thread lacked uniform thinness. Ghosh also showed his loom but it was not a finished manufacture. From this year, several sessions of the Mela were held in the districts as well as in the capital.

In the tenth session (1876), Dwijendranath Tagore presided, Somprakash (28th February 1876) reported that Babu Girish Chandra Mukhopadhaya of Andul had presented a letter press and a paper manufacturing machine. Sadharani (27th February 1876) reported Monomohon Bose's speech on the swadeshi achievements. He specially mentioned Sitanath Ghosh who had invented a cotton mill and a loom which could be run by draught animals, steam or manual labour. The first was an improvement on Charkha and produced treble the amount of yarn done by the latter. The loom was capable of producing in a single effort over twenty yards of cloth. The cloth was not yet of fine quality but in future would compete with Manchester given due encouragement.[5]

Sitanath Ghosh also lectured on Science and Technology even before he demonstrated his inventions. In fact he made his first appearance on the platform of the National Society which was created from the fourth session of the Mela in 1870, in conjunction with the Mela. Here a lecture was arranged each month. The first lecture was given by Sitanath Ghosh on machinery. He demonstrated his new air pump and powerloom. The latter could produce the work of four persons by the simple operations of one operator. On the same platform Sitanath gave his discourses on electricity. He himself says: 'In summer of the year 1871, being requested by some of my friends, I delivered two successive addresses at the National Society's Meeting in the Calcutta Training Academy's Hall on the ideas I conceived about the electrical and magnetic imports of the said practices.[6] He read a paper on the causes of infectious diseases on the Society's platform in the sixth session of the Mela in 1872. It had valuable contents which were thought worth communicating to western Scientists and the Government in English. Encouraged by such discourses, Nabagopal Mitra, later Secretary of the Hindu Mela opened the National School at the Calcutta Training Academy, at 13, Cornwallis Street on 1st April 1872. The National Paper reported the lecture by Shyama Charan Srimani and Rajkrishna Mitra from the Art School and the School of Physical Sciences respectively who took over teaching geometrical drawing and Chemistry at the school. Srimani observed, 'The study of geometrical drawing gives one the power to delineate by plans, elevation and sections the length, breadth and height or in other words the magnitude and make of any object whether a house, a bridge, a boat or a fort and as all such delineations could be made by the same principles, a knowledge of this branch of Art is of equal importance. To the Architect, the Mechanic, the enquirer (I should also mention the would be Indian CS) and in fact to all classes of mechanical workmen such as carpenters, bricklayers, blacksmith ...."[7]. Hindu Mela thus provided the impetus and the sounding board for a groping indigenous technology. It was still crude and rude, faltering and faulty. It was an unfinished revolution, more a technique than proper technology.

More about Sitanath Ghosh

From the short life-sketch of Sitanath Ghosh by Jogindranath Samaddar in Prabasi, we come to know that he was born in Roygram, Jessore in 1248 BS (1842). He was born with a big head destined to carry big ideas. He passed Entrance Examination from Narail Private School winning a scholarship. He entered Calcutta Medical College for his aptitude in scientific pursuits. However he fell ill and had to leave his studies. On recovery, he picked up Ayurveda for some time but he could not continue. He later served as a school teacher and a Government servent. He left this for its drudgery. Instead, he privately studied electricity and edited Hindu Patrika from Calcutta in which he published many scientific papers. In 1872 he was made the editor of Tattwabodhini Patrika by Devendranath Tagore having been impressed by his two lectures on Electricity and Magnetism at the National Society in 1871. In Tattwabodhini Patrika, (Nos. 351, 352, 353 and 365) he continued his writing on the same subject. Sitanath fell ill again and went home, where he started a charitable dispensary. Debandranath donated Rs. 7,000 towards the scientific work of his Bramho Protege. Sitanath invented a weaving machine which made him famous.

The Rani of Betia offered Rs. 50,000 to buy it but he did not sell it nor could he maintain it. He also invented a wheat pounding mill, a mechanical plough which could be drawn with the help of one bullock and a mechanical boat. His other inventions were writing and press ink. He gave the formula to his friend Amrita Lal Roy. A.L. Roy's ink was well known. Another invention, scientific amulate was given to Brajamohan Kar for marketing and he made a good business out of it. Sitanath founded an English School, a Library and a Post Office in his native village. He also organised an annual Agricultural and Educational Fair in Roygram. From the river front to the fair site, he set up telegraph lines for dignitaries to speak to the office from the ferry ghat for their reception. Sitanath was a local leader of the Bramho movement. Bijoy Krishna Goswami came to Roygram, visited the fair and stayed for two days for his lecture.

Sitanath's greatest scientific achievement was the Magnetic Healer,according to himself, which he described in his book on Medical Magnetism. He derived the principle from the daily prayer of the Hindoos (Anhik Mantra) and also the Vishnu Purana. The trident over temples was based on the theory of magnetism to prevent their collapse from lightning strikes. 'It has been found that', writes Sitanath in his book, 'the human body is a magnetisable object though far inferior to iron and steel'. In 1880, he started his chamber at 54, Mechhua Bazar Street and started treatment of patients by electricity and magnetism. His Magnetic Healer, one small and one big (2 ft. and 4 ft.) was composed of 6,000 and 10,00 ft. of copper wire passing through a wooden frame cushioned with mat and jute, wax and leather linings. They had brass hooks at the end of which were attached the wires of the galvanic battery for creating the magnetic field. The patient would be made to lie on this platform so that electric waves passed through the body. Sitanath claimed that "every description of indisposition known is partially or entirely removed as it is light or serious". He claimed to have cured two hundred patients in only two years but he died prematurely in 1840 at the age of 42 leaving its immense possibilities open to the posterity.[8] To this may be added the tribute from Amrita Bazar Patrika of 24th March 1870; "He had the ingenuity from his childhood to invent machines. He has already produced an air pump and an engine. He could take a patent in the first but the engine is in an experimental stage. But this is no mean achievement. Instead of steam, air pressure is monitered to move the machine. He has given many lectures on science and technology at the National Society and in the open sessions of the Hindu Mela. He is a pioneer of science and particularly in technology in our country. He has written many articles in Tattwabodhini Patrika based on his scientific research.[9] Even Rabindranath reminisces about Sitanath in his memories. As a child he was immensely delighted by the various scientific experiments carried out by Sitanath in the Jorasanko house on Sundays to teach him elementary science.[10]

In Theory

Sitanath in his work, Medical Magnetism describes himself as the founder of Electropathy or the Magnetic system of Treatment in India. The earlier research on this subject was reported by him in the columns of Tattwabodhini Patrika on Electricity and Magnetism written in 1872. Like P.C. Roy of later years, Sitanath Ghosh also traces the root of this science to our ancient past when sages like Sankhya, Aryabhatta, Sankara Charya and Bhaskaracharya propounded the principles of science. They dealt with even all kinds of material sciences including physiology though they were frequently blended with daily rituals. From these treatises one can reconstruct their ideas of the properties of Electricity and Magnetism which strikingly resembled the European notions of electricity like positive and negative currents were described as masculine and feminine forms. Then there is an elaborate description of conductivity through different materials high or low which could be manipulated to produce energy for human use. He elaborates on the variety of conductors through which the heat and sound could be produced at will and their velocity could be utilised for various purposes for physiotherapy. Sitanath also discussed the properties of magnetism with authority particularly on the magnetic poles and their behaviour. He further dilates on the magnetic properties on the earth and its minerals and the human use to which they could be put. Sitanath then tries to reconcile European science with our ancient scientific knowledge. His example is the trident of the temple which absorbs the lightning which could strike and destroy the temple and gives him reasons why. His explanations reveal his thorough grasp of physics of electricity and magnetism. In the following articles, he established the connection between, the electricity and magnetism outside and those within the body and their interactions particularly on the nervous system. The ancient rituals guiding our food, dress and daily chores all had scientific bases. They controlled and decontrolled the impact of electricity and magnetism on the body. Role of the metals we use and other materials were all determined by the impressive theory of magnetivity. This led Sitanath to apply the laws of physics to physiotherapy where the lying position becomes supremely important. It has to coincide with the magnetic poles of the earth for advantage. He finds absolute justification for our ancient system of sleeping with the head turned towards the south and feet to the north or east and west. Having stated his principles, he turns towards what he later called medical magnetism or electropathy as a new theory of medical science of which he was the pioneer in India if not the world.[11]

Sitanath did not rest contant with theory alone. As was his wont, he soon invented an apparatus to demonstrate what he formulated. The outcome was the invention of the Medical Healers about which we have written already. He was theoretically and practically a complete scientist or more correctly a true technologist.

In Search of Technology

We have concluded in the foregoing section that technology was technique informed by science, a step higher than what an artisan or mechanic could do by fabrication, improvisation and innovation. It aims at discovering the principles which could generate patents and further fabrication without difficulty. It is true that techniques or craft did emerge earlier than scientific theories but there is no reason to treat them autonomously in the history of science and technology. They were infact symbiotic. One aided the other in their evolution upto the present times in human history. While technique showed how to do it, science showed why it should be so. Once the scientific principles were known, there could be improvement of art and craft. Both combined to produce technology. While a Wyatt or Hargrteaves or a Davy paved the way of Industrial Revolution in England. It was brought to its second stage by scientists like Newton, Faraday and Rutherford. To say technique had its own primacy and science was an attempt to Sanskritise it or that the former was the art of the masses and the latter that of the elites is to create sociological complexities to obfuscate the complementary of the two. As J.J. Saloman rightly points out, the uses of the word, technology in English confounds it with what is known in the continent as technique where technology means much more specialised and more advanced stages of technique. To redress the inadequacy of the word technology, the English have coined the expression high technology to mean techniques of above average levels. In trying to sum up the debate, Saloman concludes : 'If in theory, science is on the side of pure reasoning, technology is on the side of utilitarianism; however; its efficacy is still derived from scientific rationlism ...... a non existent science : technology as an independent discipline, the subject of research and teaching has remained a project without a future.[12]

If the continental meaning of technology is accepted, Sitanath Ghosh was undoubtedly a technologist of a high order. He combined both theory and practice and therefore was a pioneer in this field in Bengal. The debate over science versus technology during the foundation of the Indian Association for the Cultivation of Science by Dr. Mahendra Lal Sircar in 1876 between the proposers of IACS and the champions of a mechanics institute was laid to rest by stalwarts like Rajendra Lal Mitra and Father Lafont when they stressed the need for advancement of science for the improvement of artisanal skills of our smiths, potters and carpenters. They were in fact looking forward to technology of which Sitanath Ghosh was the contemporary exemplar and father figure.[13] The efforts of Mahendralal and Sitanath found fulfilment in the activities of the Dawn Society and the Bengal Technical College.

References

1. Jogesh Chandra Bagal Hindu Melar Itibritta (Bengali) Calcutta, 1967, Appendix I, p. 91.

2. Soumyendranath Tagore 'Evolution of Swadeshi Thought' in A. C. Gupta edited, Studies in Bengal Renaissance, Calcutta, 1977 pp. 209-10.
3. C. Palit, New View points on Nineteenth Century Bengal, Calcutta, 1980, p. 43.
4. *Ibid.*, p. 44.
5. Jogesh Chandra Bagal, *op.cit.* passim.
6. *Ibid.*, p. 27, quoting Sitanath Ghosh, Medical Magnetism, p. 9.
7. *Ibid.*, p. 84, Section on National School, quoting the National Paper, September 4-1872.
8. Jogindra Nath Samaddar, 'Sitanath Ghosh' in a Prabashi 1319 BS (1930), passim.
9. Jogesh Bangal, *op.cit.*, p. 61, Quoting Amrita Bazar Patrika, 24 March, 1870.
10. Rabindra Nath Tagore, Jiban Smriti in Complete works, Centenary edition, West Bengal Government, Calcutta, 1962, pp. 27-22, cf : Rabindranath mentions Sitanath as Dutta and not Ghosh because of failing memory. He was none other than Ghosh himself as circumstantial evidence points out. See Prasanta Kumar Pal, Rabijibani, Vol. I for a second opinion, Calcutta, 1982-83, pp. 134-35.
11. Tattwabodhini Patrika, Agrahayan 1794 SK (1872), and Poush, Nos. 351, 352, 353 and 365, cf : Sitanath's Statement in Medical Magnetism with reference to these articles as quoted in Samaddar *op.cit.*: 'The essays were illustrated by four engraved plated viz. 1. A temple with an iron trident. 2. A naked man with a long trifurcated iron bar in his right hand and a buffalow horn bugle on his left shoulder making in fact the picture of a silary or hailstone preventer. 3. An asthmatic patient with a

maduli. 4. A man lying down on the surface of the northern sphere of the earth with his head placed southward.

The singularity of the explanations combined with the oddness of the plates excited, as I learned, laughter and ridicule among the ordinary readers and applause and admiration mingled with doubts among the more intellegent class of readers original in Bengali can be quoted in extract.

মনুষ্য প্রভৃতি যাবতীয় জীবের শরীরে সর্ব্বসময় কিয়দ পরিমাণ মুক্ত তড়িৎ বিদ্যমান থাকে। এই তড়িৎ পরিপাক যন্ত্রাদির বিবিধ রাসায়নিক প্রক্রিয়া হইতে উদ্ভূত হয় এবং শরীরে অবস্থিতি করিয়া নানা প্রকার কার্যসাধন করে। ঐ তড়িৎ হইতেই শরীরস্থ স্নায়ু মন্ডল বলপ্রাপ্ত হয় এবং সেই বল দ্বারাই শরীরের ও মনের সমুদায় কার্য্য প্রকাশিত হইয়া থাকে। পরীক্ষা দ্বারা প্রতীত হইয়াছে যে, শরীরে তড়িতের পরিমাণ অল্প হইলে স্নায়ুমন্ডল ক্ষীণ হইয়া পড়ে। সুতরাং তখন তাহার দ্বারা কোন গুরুতর কার্য্য সাধন করা যায় না। আবার তড়িতের পরিমাণ অধিক হইলে, স্নায়ুমন্ডল বলিষ্ঠ হইয়া ওঠে। তখন তাহার দ্বারা বিস্তর বলসাধ্য কার্য্য সাধন করা যাইতে পারে।

12. Sabyasachi Bhattacharya and Pietro Redondieds 'Techniques to Technology', New Delhi, 1990, forward and pp. 242-245, wrote by J.J. Saloman.

13. C. Palit, 'Mahendralal Sircar and the Quest for National Science' in D. Kumar edited Science and Empire, New Delhi, passim.

14. তদেব দ্রঃ ষষ্ঠ অধ্যায়

15. হরিদাস মুখোপাধ্যায় ও উমা মুখোপাধ্যায়, 'জাতীয় আন্দোলনে সতীশচন্দ্র মুখোপাধ্যায়,' কলকাতা ১৯৬০ "Origins of National Education Movement," Calcutta, 1957, Section on Dawn Society.

16. 'The Dawn' Vol. V, No. 2, Sept. 1901.

17. *Ibid.*, Vol. V, No. 4, Nov. 1901.

18. *Ibid.*, Vol. VI, No. I, Aug. 1902.

6

Swadeshi Technical Education: Bengal Technical Institute (1906-1928)

Introduction

Rajendralal Mitra was the pioneer of national technical education in nineteenth century Bengal. His aim was to impart technical education to novices in the institute founded by him in the sixties of the century. Though the institute provided the minimum training in technology to help them earn a living, technology as a subject did not have any advancement here. Rajendralal was well aware of it. This is clear from his speech in support of the founding of the Indian Association for the Cultivation of Science in 1876.[1] When Dr. Mahendralal Sircar wanted to establish the Association, he was opposed by the leaders of the Indian League. They wanted to establish a technical institute instead for unemployed and semi-literate people to enable them to earn a living. This led to a historic debate in the Town Hall under the chairmanship of Lieutenant Governor Richard Temple. The Indian League sought official patronage as a newly born political party and wanted to scuttle Mahendralal's National Science Association by proposing a technical institute to be called Albert Institute in commemoration of the impending visit of the Prince Consort to India.[2]

Rajendralal Mitra and Father Lafont supported Mahendralal's association vigorously. Rajendralal Mitra pointed out that without cultivation of science, technical education did not become technology. It remained at the level of artisans.[3] The house finally voted in favour of the I.A.C.S. and it was born in 1876. But for political motives, the proposed technical institute of the Indian League could have been a full fledged technical institute as can be found from Dinanath Sen's report in the Hindu Patriot.[4] It was felt in the Indian circle that a national science association was needed under Indian supervision for national cultivation. The technical institute would have been dependent on foreign instructors. The British government had no interest in promoting the I.A.C.S. The I.A.C.S. was the first nationalist body of its kind which was dedicated to create a pool of scientists who would also supervise technical education if necessary.[5]

The idea of swadeshi science and technology took hold of the mind of the educated Bengalis from the days of the Hindu Mela. A technical fair was regularly held as a part of the Hindu Mela since 1867. The art and craft of artisans and budding scientists were regularly displayed there. Among the intellectuals Rajnarain Bose, Bholanath Chandra and Manmohan Bose were the chief patrons of the Hindu Mela and they laid great emphasis on all kinds of swadeshi enterprise. The inventor who caught every eye for his very original inventions was Sitanath Ghosh. These included a sewing machine, an air pump, a mechanical plough, telegraph communication, of short length and a magnetic healer. Sitanath used to regard the Healer as his greatest invention. It cured two hundred patients of various nervous diseases. Sitanath was also well known for his scientific articles published in the Tattawatbodhini Patrika. His article on electromagnetism entitled him to be regarded as the first Bengali technologist in theory as well.[6]

The next important centre for technology after the era of the Hindu Mela was the Dawn Society (1902-1906) founded by Satish Chandra Mukherjee. Even before it, Satish Chandra had started the *Dawn Magazine* (1897) in which articles of science and technology were regularly published. The *Dawn Magazine* and Dawn Society continued the tradition of popular science started by Dr. Mahendralal Sircar in the series of scientific and technological lectures at the I.A.C.S.[7] Swadeshi technology was specially discussed in the columns of the *Dawn Magazine*. Radhakamal Mukherjee, Rabindranarayan Bose, Binoy Sarkar, Kunjabehari Sen and Satishchandra himself wrote these articles. The *Dawn Magazine* carried more than forty articles on science and technology. Kunjabehari Sen had been to Japan. He gave demonstration lecture on textile manufacture at the Dawn Society and organized sale of cotton piece goods produced by the apprentices of the Dawn Society in a shop in Burrabazar. The Dawn Society had regular courses in industrial and technical education at both primary and higher levels. It equiped the apprentices in day to day mechanical work which helped them to earn a living. Thus the Dawn Society provided the immediate background to the National Council of Education and its technical school, the Bengal Technical Institute or the B.T.I.[8]

The Birth of BTI

The political whirlpool caused by the University Act of 1902 and the Partition of Bengal (1905) transformed an educational society like the Dawn Society overnight into the giant body of the National Council of Education in 1906. The University Act had sought to disaffiliate private colleges set up by the Bengali middle class for their nationalist activities and the Partition was a bigger assault on the same class to break their solidarity. These acts, instead of curbing nationalism intensified it. In the first phase it aimed at national reconstruction and its chief activity was to establish an alternative nationalist system of education. The National Council of Education was born on 1st June, 1906 with Rashbehari Ghosh as the president and Asutosh Chowdhury and Hirendranath Dutta as joint secretaries. Surendranath

Banerjee hailed it as "the first great constructive effort of the Swadeshi movement".[9] To start with, the NCE launched two institutions, the Bengal National College for general science and arts and the Bengal Technical Institute for technical education. Binoy Kumar Sarkar, a teacher of the Council of Education wanted it to function as a full fledged university and had the ambition to impart literary, scientific as well as technical education from the lowest (primary) to the post graduate standard."[10] This three stream education was not approved by some NCE members like Taraknath Palit, Nilratan Sarkar, Bhupendranath Bose, Rajendranath Mukherjee and Manindra Chandra Nandy. They were mainly for technical education. Taraknath Palit was instrumental in the establishment of a Society for the Promotion of Technical Education under the NCE and he got a bill passed by its executive committee forming the SPTE on the same date in 1906. The SPTE founded the Bengal Technical Institute or the BTI.[11] In a report issued in July 1906, the purpose of the BTI was thus explained : "Our attention is first directed to the great industrial awakening that has of late taken place in the country. Everywhere signs of this new industrial spirit have manifested themselves. Our eyes have been opened to the vast natural resources of the country lying practically undeveloped. The success of Europe in the field of industrial enterprise has also been an eye opener to us and has stimulated our energy".[12]

The BTI was opened on 25th July 1906 at 82, Upper Circular Road with Pramathanath Bose as its first principal. Rashbehari Ghosh was the president of both NCE and SPTE. He was also the president of the executive committee of BTI. At the initial stage, the BTI introduced two kinds of courses. One was a three-year intermediate course and the other was a four year secondary course. The non-matric students could also get admitted to the intermediate course. They mainly learnt fitting, dying, carpentry, eletroplating, lithography, soap-making and tanning. But the course included general studies in Physics, Chemistry, Mathematics and English. Those who had passed at least Matric or equivalent examination were admitted to the intermediate course. The main three subjects here were (1) Mechanical and Eletrical Engineering; (2) Applied Chemistry; (3) Geology. In the first year of this course, instruction were given in Physics, Chemistry, Mathematics, English and Drawing.[13] Some of its illustrious teachers were Pramatha Bose, Sarat Datta, R. Cowlon, Prafulla Mitra, Bhopal Sen, Jogesh Ghosh, Nibaran Ray, Krishna Samaddar, Satya Deb and Jatin Das Gupta.[14] The BTI had its manufacturing unit which produced new industrial items and also undertook repair work. In 1908 it had 144 students in the two courses. By 1910 the roll strength was 144. The reason for it was explained by Hemchandra Guha, a student of BTI and later Vice Chancellor of Jadavpur University in this way. In those days people understood by a career, professions like advocate, clerk, police official, broker, businessman, doctor etc. Engineering was not thought of as a career.[15] Besides B.T.I. had no hostel and job prospects for trainees of this national institute were not bright.

Bengal National College Versus BTI

Both these colleges were established almost about the same time. Both were under the National Council of Education and Rashbehari Ghosh was the president of both. Yet there was a lot of dispute among their founders about the courses to be offered for national reconstruction. The National Council of Education wanted to establish a parallel university with scope for studies in Arts, Science, Commerce and Engineering. But the BTI only wanted to impart technical education. Besides NCE was influenced by nationalist thoughts opposing official educational policy whereas BTI did not want to oppose Government. Thus the BTI was called moderate, loyalist and pro-government by the stalwarts of the NCE. The celebrated teacher of the Bengal National College, Binoy Sarkar has described this difference in the attitude of the two bodies as 'Kultur kamph' or 'culture war' in his history of the NCE.[16] In his words, "this polarity of views conveniently boiled itself down into the anti-thesis : nationalism versus materialism and culture versus crude mistrification".[17]

Despite this conflict, the two bodies amalgamated into one institute under the NCE on 25th May, 1910. The long standing dispute ended and the council declared its intention to "establish and maintain a central educational institution for imparting general and techical education in all its phases according to the scheme of studies approved by the Council."[18] But the crisis remained. The National College had neither students nor money but the BTI had money and few students. A situation drifted like this till the annulment of partition in 1911. Because of low salary, many of its reputed teachers left the institution. Among them Pramathanath Bose, Jagadish Chandra Bose, Sarat Kumar Datta and Jatindra Kumar Dasgupta left for government jobs at higher pay.[19] The crisis was aggravated in 1912 when barrister T.N. Palit and Maharaja of Cooch Bihar decided to withdraw their money from the B.T.I. T.N. Palit gave away his bequest including the house at 82, Upper Circular Road to Calcutta University. Calcutta University under Sir Asutosh Mukherjee pursued a curriculum which was approved by T.N. Palit and others. But BTI being a nationalist institution could not come under Government supervision and somehow carried out its activities from Panchavati Villa on Canal East Road in Maniktala. [20] Some of its old teachers like G.C. Sen, Professor Cowlon, Sushil Chakraborty and Manindranath Banerjee left it in 1913. The salary of teachers of the Council was very low and irregular. From an account of 1914 it appears that a teacher of the engineering college had a monthly salary of Rs. 250, teachers of Physics and Chemistry got only Rs. 100/-, a teacher of English got Rs. 75/- and the demonstrators got only Rs. 40/-, Binoy Sarkar observes, "the self sacrifice and the economic martyrdom which some of the members of the Bengal intelligentsia were prepared to undergo were in the interest of an educational transformation and technoscientific remarking"[21]. To this Amitabha Mukherjee adds, "those who could not be persuaded to undergo this self sacrifice and economic martyrdom naturally left their job after sometime in search of a more remunerative employment."[22]

During this hour of crisis, Sir Rashbehari Ghosh donated over 13 lakhs of Rs. to the NCE. Others who also donated were Manindra Chandra Nandy, Asutosh Chowdhury and Byomkesh Chakraborty.[23] This munificence solved the fund crunch of the NCE to some extent.

Since 1910 the student enrolment of the BTI improved. But the National College offering humanities failed to draw students. Between 1921-25 the BTI became a flouishing institute whereas the National College was closed down in 1917. Binoy Sarkar remarks, "For the course of a few years, be it said at once, Bengal National College found itself with empty benches and automatically ceased to exist. By 1916-17 not a student cared to come for literary and scientific instruction on national lines. The institution that endured and survived was the institution of alleged materialism or bread and butter philosophy, Bengal Technical Institute."[24]

From BTI to the College of Engineering

During the World War I, many nationalist leaders decided in favour of supporting the British authorities for gaining political concessions after the war. This change of attitude is distraught the activities of the NCE. Some eminent teachers, like Bhim Chandra Chatterjee left the B.T.I. in 1916. B.T.I. still maintained its programme of technical education. A new executive committee was formed in 1918 to expand the activities of the B.T.I. It was decided to buy land in the suburb for its new campus, encourage students to find self employment, grant, scholarship for research, build hostels for residential students, organise sports and institute awards for general proficiency and sports. The most important step was to employ foreign returned Indian students as teachers of the B.T.I. One such new recruit was Professor Hemachandra Dasgupta, who returned after his higher studies in Michigan, USA in 1918. As Binoy Sarkar observes, the period 1910-18 was "a period of struggle and experimentation in the history of BTI".[25] But its survival was "a specimen of self sacrifice, creative patriotism and strenuous exertions."[26] The next decade was a period of resurgence of the B.T.I.[27] The anti-Rowlatt agitation (1919) and the start of the non co-operation movement under Mahatma Gandhi (1920) brought back its nationalist agenda. The student enrolment of the BTI rapidly increased. In 1919 the number was 150, in 1920 it was 247 and in 1921 it swelled to 520. It became necessary to find a new campus and after hectic search a plot of 100 bighas was taken on a lease at rent of Rs. 210 per year for 99 years from the Calcutta Corporation at Jadavpur. The foundation stone of the Aurobindo building was laid on 11th March 1922 and the building was completed in 1924. The BTI was then shifted to this new building from Panchavati Villa in Maniktala. This was now its permanent address. Within 5/6 years this campus had its laboratories, workshops, boys' hostel, power house, staff quarters etc.[28]

From 1921 it was decided to expand the scope of studies in engineering. A Syllabus Revision Committee was formed of which Sir P.C. Roy was the chairman. The new curriculum included Chemical Engineering besides Civil and Mechanical. It was also proposed to open a school of Agriculture and a Medical College. But they did not materialise.[29] It was also decided to improve the faculty by sending some of the teachers abroad for training. The first three trainees were Sri Hiralal Ray (Chemical Engineering); Sri Satish Chandra Bhattacharjee and Sri Jotindranath Bose (Mechanical Engineering). They came back and rejoined the BTI in 1928 after higher studies in Germany. Others who came back from abroad and joined the B.T.I. were Baneswar Das, Sudhir Chakraborty, Hemchandra Guha and Triguna sen.[30] Hemchandra Guha stood first in first class in Electrical Engineering from Edinburgh University. His sense of joy has been recorded in his reminiscences.[31] They all adorned the staff of the BTI. It was now felt that the BTI should be renamed as the College of Engineering and Technology as a parallel institution to the Shibpur B.E. College. There was no turning back since then. By 1925 the CET was turning out nearly 100 qualified engineers for the cause of the country.[32] It was recognised as a major engineering college both at home and abroad. The apathy of British authorities could not deter its success. In 1956 when Jadavpur University was founded, the CET became its integral part.

Thus the BTI which was a small industrial school in 1906, braved all crisis to become an engineering university. Binoy Sarkar describes how his three stages of technical education viz. Mistrification, Modernisation and Industrialisation were all visible in the growth and development of the BTI into the CET.[33] This illustrates the transition from technique to technology. At the level of BTI, it was only a Mechanics Institute which Sarkar called Mistrification.[34] When Science supported it, it got elevated into technology for industrialisation. Its contribution to national reconstruction is immense. The Shibpur Engineering College had Civil and Electrical departments and provided personnel for mainly Government departments whereas the CET taught many other branches, like mechanical and chemical engineering and produced not less than 500 entrepreneurs like Alamohan Das before Independence made their name all over the world.

References

1. Samarendra Nath Sen : Vijnanyacharya Dr. Mahendralal Sircar, Calcutta, 1980, p. 38 also *A Century* (Centenary Volume of Indian Association for the Cultivation of Science) Calcutta, 1997, see relevant chapters.

2. Deepak Kumar (ed.) Science and Empire, New Delhi, 1985 see the article 'Dr. Mahendralal Sircar and National Science' by C. Palit.

3. Siddartha Ghosh : Kaler Shahar Kalkata, Calcutta, 1993 also C. Palit's 'Mahendralal Sircar O Gananukhi Vijnan'

(in Bengali) in Aparajita Basu (ed.) Vijnan O Prajuktir Itihas, Calcutta, 1997.

4. Siddhartha Ghosh's unpublished monograph 'Rajendralal Mitra".

5. A Century, *op.cit.*, see introduction.

6. Tattawabodhini Patrika (in Bengali), Aswin, 1789 (Shakabda) 398, p. 227 also Jogesh Chandra Bagal, Jatiyatar Naba mantra ba Hindumelar Itibritta (in Bengali) Calcutta, 1352 (B.S.) p. 25.

7. Haridas and Uma Mukhopadhyay : Jatiya Andolane, Satish Chandra Mukhopadhyay (in Bengali), Calcutta, 1960 passim also Subrata Pahari; Satish Chandra Mukherjee and National Education (1865-1948) unpublished M.Phil dissertation, passim.

8. *The Dawn*, Calcutta 1906, relevant vols.

9. Amitabha Mukherjee : Fifty years of National Council of Education, Calcutta 1992, p. 36.

10. Benoy Kumar Sarkar : Education for Industrialisation Calcutta, 1946, pp. 4-96.

11. Amitabha Mukherjee : *op.cit.*, p. 39.

12. Binoy Kumar Sarkar. *op.cit.*, p. 91.

13. Amitabha Mukherjee, *op.cit.*, p. 112.

14. Binoy Kumar Sarkar, *op.cit.*, pp. 88-99.

15. Hemchandra Guha : Man O Manisha (in Bengali) recorded by Ashim Mukhopadhyay, Calcutta, 1992, p. 42.

16. Binoy Kumar Sarkar, *op.cit.*, pp. 94-97.

17. *Ibid.*, also for further details, Haridas and Uma Mukherjee : Origins of the National Education Movement, Calcutta, 1957 session.

18. Annual report of National Council of Education for 1909-10, p. 2.
19. Amitabha Mukherje, *op.cit.*, p. 115.
20. *Ibid.*, also N.K. Sinha : Sir Asutosh Mukherjee : A Biography.
21. Binoy Sarkar : *op.cit.*, pp. 117, 119.
22. Amitabha Mukherjee, *op.cit.*, p. 116.
23. *Ibid.*, pp. 114-15.
24. Binoy Kumar Sarkar, *op.cit.*, p. 112.
25. Amitabh Mukherjee, *op.cit.*, p. 118.
26. Binoy Kumar Sarkar, *op.cit.*, see relevant chapters.
27. Minute Book of National Council of Education, 1906-24 passim.
28. The Journal of the College and Engineering and Technology, Bengal, 1938, p. 36.
29. Minute Book of the National Council of Education 12th December 1918 also T. K. Roy, (ed.) Life and Times of Hiralal Roy, Calcutta, p. 139.
30. Amitabha Mukherjee, *op.cit.*, pp. 123-24.
31. Hemchandra Guha *op.cit.*
32. Binoy Kumar Sarkar *op.cit.*
33. *Ibid.*
34. *Ibid.*

7

Binoy Kumar Sarkar : An Ideologue of Swadeshi Industrialization

Binoy Kumar is not a very well known personality and writer to the reading public nor to the scholars of the freedom struggle. But to the intellectuals of pre independent Bengal, he was known to be the ideologue of swadeshi industrialisation and father of Indian sociology. He had encyclopaedic knowledge in most of the social sciences and was a prolific author. He remained a mentor of the Bengalee intellegentsia since 1906 till his death in 1949.

Life

Binoy Kumar Sarkar was born on 26th December 1887 in Maldah, West Bengal. His father was a clerk in a local collectorate. His family hailed from Bikrampur, Dhaka. He stood first in the entrance examination of the Calcutta University in 1901 and entered Presidency College for higher studies. In 1903 he passed his F.A. Examination standing 9th in order of merit. He joined the B.A. class of the college with honours in both English and History and in 1905 he came out first in first class and became the Ishan scholar of Calcutta University that year. He did not join M.A. class but at the request of his friends he appeared in the M.A. examination in English and got second class. He did not go for the deputy magistrate's job instead, he took the plunge in the swadeshi movement which was raging in Bengal.[1] While in the Hindu Hostel, he had a close friendship with his fellow boarder Radha Kumud Mukherjee. Both of them became disciples of Satis Chandra Mukherjee and joined his Dawn Society, which was active from 1902 and was dedicated to the cause of swadeshi. He became involved in National Education Movement for which he was already working on the platform of the Dawn Society since 1904.[2]

In his later life, Binoy Sarkar in his reminiscences fondly recalled the intellectual atmosphere of the Dawn Society and the great patriotism and self sacrifice of Satish Mukherjee as its pivot. The Dawn Society was dedicated to nation building activities and it started the swadeshi movement in 1903-4 before the actual Bengal Revolution broke out in 1905-6. Satis Mukherjee and his associates became the propagators of the swadeshi movement in the town and the country much before 1905.[3] The members of the Dawn Society were already initiated in swadeshi by Satis Mukherjee who was the father of the Swadeshi movement and Binoy Kumar Sarkar was one of the trusted disciples. He was a core member with Rabindra Narayan Ghosh, Radhakumud Mukherjee and Haran Chandra Chakladar. Later Nilratan Sarkar, Jagadish Chandra and even Rabindra Nath joined the activities of the Dawn Society. Satish Mukherjee himself and the core members incessantly wrote in the Dawn Magazine in favour of National Education. They provided the nationalist ideology to the later swadeshi movement. Satish Mukherjee himself wrote on education including scientific education, Radhakumud wrote on history, Rabindra Narayan Ghosh on economics, Haran Chandra Chakladar on history and sociology and Binoy Sarkar on economics, history and sociology. Binoy Sarkar became the prophet of economic swadeshi and became actively engaged in founding national schools in the districts of Bengal and teaching in the Bengal National College.[4]

He remained the honorary teacher of the National School and College till 1914. From 1914 to 1925 he was abroad visiting U.S.A, Africa, Europe and Asian countries and giving public lectures on India. He returned to India in 1926 and became a professor of economics in Calcutta University, which post he held till his death in 1945.[5]

Binoy Kumar Sarkar has an enormous corpus of writing in social science and literature to his credit but was known to be a specialist in economics and sociology. He was the author of such famous titles as—The Positive background of Hindu Sociology, A Structure of Hindu State, Education for Industrialisation, A scheme for Economic Development, Prosperity and Depression. The purpose of the essay is however to highlight his thoughts on economic swadeshi as he was known to be an ideologue of swadeshi industralisation.

The Ideologue of Economic Swadeshi

Binoy Kumar Sarkar in his reminiscences tells us that he constantly thought of the best and most urgent means to serve the country. He thought of economic recovery as the main goal and wanted to orient the minds of the educated classes towards swadeshi enterprise.[6] Satis Mukherjee was the pioneer in this field. At the Dawn society, he instructed the basic principles of economics rigorously so that the Bengalee youth was inspired to undertake swadeshi enterprise for economic regeneration of the country.[7] For practical training, the Dawn Society opened a swadeshi shop in 1903-04. The members of the society went for manufacture of swadeshi goods which ranged from cotton piece goods, oil, soap, furniture to small tools and spare parts of machinery. Even salt and sugar merchandise were displayed in the swadeshi shop. This was the precursor to later swadeshi enterprise undertaken by the students of the Bengal Technical Institute and National College under the National Council of Education.

The swadeshi shop of the Dawn society therefore, played a pioneering role in promoting economic swadeshi. Sarkar laments that in the swadeshi period (1905-10) a real entrepreneur and businessman was rare among the Bengalees. There was none who could become an ideologue of swadeshi industrialisation. Leaders like Satish Chandra Mukherjee, Rabindra Nath, Brahmabandhab Upadhyay were mentors in the cultural field. Bipin Chandra and Surendra Nath Banerjee were political leaders of anti-colonial struggle. There was void in the economic field and there was no expert who could guide the youth in the business enterprise and setting up of factories and machinery. He recalls only one Ramakanta Roy of Sylhet who was trained in Japan in engineering and was a champion of swadeshi enterprise. Under these circumstances, Satis Chandra was the only guide who provide economic ideas to the Bengalee yputh. The other inspiration was Ramesh Chandra Dutta's Economic History of India, which was like Bible during swadeshi period, in which colonial rule was blamed for all sufferings and economic decline. Among the landed aristocrats there were two classes, one group was actively connected with swaraj and boycott but the other group was more inclined towards swadeshi enterprise and business line. They were in business for two/three generations and dreamt of an industrial-revolution in our country. Roys of Bhagyakul, Nandis of Kasimbazar and Lahas of Calcutta were among them. In this milieu, Binoy Kumar Sarkar took upon himself the mantle of the economic messiah of the country.[8]

The main currents of his swadeshi thought are contained in his history of the National Council of Education entitled, Education for Industrialisation. In the preface to the book, Sarkar wrote.

Although historical it does not deal much with the past. I am devoting more thought and time as well as space to the present than to the bygones. This is manifest in the attention that is being bestowed on 1945 and even the few weeks of 1946. Nay, I am interested more in tomorrow and day after tomorrow than in today. My obsession is futuristic. No antiquarian or archaco bias can be expected in such an analysis.[9]

It deals with the vishva-shakti (world-forces) in the engineering and technological education of diverse countries, and exhibits Jadavpur in the perspective of international pedagogies. It seeks, further, to invite attention to the world's progress in industrialism, technology, industrial research, scientific discoveries, etc. Nationalist India is thereby being compelled to ponder over the tremendous lags and backwardness of the Indian economy vis-a-vis the economic developments of go-ahead peoples.[10]

He clearly defines the goal of this work in terms as follows:

At every step, again and again, my interest lies in exploring the possibilities of Jadavpur College as an instrument for the further "mistrification", industrialization, 'factorification' and economic modernization of the country in the near future. The survey is meant to be a help to professional engineers, technologists, business managers, and industrial researchers both as individuals as well as organizers, directors or promoters of factories, workshops and research institutes.[11]

Sarkar gives an early indication of his technological formulations in the preface itself:

It is perhaps not irrelevant to say at the outset that a simple pedagogic idea runs through this study. My sutra (formula) in the pedagogies of technocracy and machinism runs to the effect that not every mistri or technician can or will become engineer. But on the other hand, nobody can become an engineer who has not been a mistri. The daily handling of tools, implements and machineries belongs to the mistri's routine and daily tool handling is the fundamental schooling for the engineer.

Further, the real or substantial training ground for the engineer is the manufacturing establishment of the market rather than the laboratory. Workshop or class room of the college. This ideology about the role of the mistri and the factory in engineering and technology, the factorification of the students of engineering and technology and the mistrification of general culture has not been intensified by extensive travels and investigations in foreign countries as well as contacts with engineers, chemists, businessmen, industrialists, both at home and abroad.[12]

Sarkar coined the word 'Mistrification' in a derogatory sense to mean elementary technical education for bread and butter and was forever fighting the trend of his time for its upgradation with technology combined with culture. Some early indications of this inner conflict in Sarkar is in the following statement.

It is worthwhile to add that Jadavpur College, while functioning as an instrument of industrialization, mistrification, technocratization, and economic modernization seeks at the same time to provide the 'humanities' also as an integral part of engineering and technological education. World history is a compulsory subject for every student of mechanical, electrical and chemical engineering. It is taught in the second year class as a weekly two hour course. Not less than fifty lectures are delivered in this subject. In the group of humanities may be listed also topics such as engineering, economics (already referred to), industrial psychology, labour economics and social work, "man the known, and man the unknown" English language and literature, geology, and last but not least, religion. The Jadavpurians, engineers and technologists, mistrified as they have been have to grow up as men of culture.[13]

To him religion was a part of this cultural orientation for man-making. He heartily applauds the practice of religious instruction imparted in the Jadavpur Campus where discourses on the Gita, Manusamhita, Bhaktiyoga, Nityachara and Upanishada were regular subjects of discourse.

He approvingly writes :

It was profound statemanship on the part of Sj. Brajen Roy Chowdhury in discussion with Satis Mukherjee, already in 1905-07, to insist on religious education as an essential limb in the policy of the National Council and to earmark a substantial fund for this purpose. Not many students perhaps attend these classes or lectures systematically. But many of those who occasionally attend them come away often with permanent lessons for the development of morale and personality.[14]

He was in favour of educational visits to industrial concerns. He writes:

Visit to industrial concerns like Basanti Cotton Mills, Calcutta Tramways Works, Bengal Belting Works, National Iron Works and Bengal Electrical Lamp Works-all in and around Calcutta-are mentioned for 1943. The Lac Research Institute at Ranchi, the Mica Mines at Kodarma, and the Cotton Mills at Gaya as well as the Hudru Falls at Hazaribagh have commanded the attention of some of these excursionists. The tech 1944 makes it a point of observe. The larger industries of course refused us admission rightway; "that is an item that remains to be straightened out between the Jadavpur College authorities and 'Big Business"[15]

Sarkar is appreciative of the theoretical writings of Jadavpur students in the journal of the College of Engineering and Technology, which took in 1928 from the Bengal Technical Institute founded in 1906. During Sarkar's young days industrial ideologies were not found but thanks to the curriculum of National Council of Education and teaching of the dedicated teachers of science and technology. By 1945 the College Journal could boast of articles like;

The Energy Index of India 'Hydroelectricity in India', 'use of synthetic Materials as Bearings for Mills and other purposes', 'vehicles driven by Producer Gas' 'beware of the Static', 'social Aspects of Engineering' 'Outline of Broad Casting Systems', 'High Pressure Technique' and 'Water Cooling System of a Motor Car Engine'.[16]

Sarkar narrates the history of the birth of National Council of Education in the stormy days of the Bengal partition and Curzon's Education policy to restrict education to Bengali Bhadraloks. The British Government was totally against the spread of technical education and confined it to the one and only Bengal Engineering College at Shibpur. National Education Movement drew its inspiration and euphoria from the anti partition movement. But its main target was to have a parallel system of national education contrasted with colonial education under Calcutta University. These being the thrust, Sarkar's history is tinged with such ideological observation as follows:

Nine of these seventeen men were lawyers. Two were medical practitioners. One was a zamindar (Land holder). The others were men of general culture. Nobody had any experience of industry, engineering, commerce or business of any sort. The National Council of Education wanted to function as a full-fledged University and had the ambition of imparting literary, scientific as well as technical education from the lowest (primary) to the postgraduate standard. But not a single engineer, industrialist or business organizer was called upon to function on the Executive Committee of this council which was established as a protest against the over literary, all too academic, unscientific and unindustrial university of Calcutta. The sociocultural conditions of Bengal towards the beginning of the present century were indeed unspeakably innocent of science, industry, engineering, technocracy, machinism, commerce, banking and business organisation.[17]

From the viewpoint of 1945, Jadavpurians ought but to admire the enthusiasm and sincerity of those leaders of the Bengali revolution (1905), who without the least knowledge and experience of engineering and technology embarked upon a venture of educational reconstruction based upon and calculated to promote industrialisation, "mistrification and economic modernization. Be it observed that these seventeen men were some of the greatest intellectuals and sincerest patriots of the time. Hardly anybody has by character and strenuous exertions been a greater inspirer, idealist and epoch-maker in politics than Aurobindo and none a greater philanthropist and educational benefactor than Rashbehari Ghosh.[18]

Commenting on the Bengal Technical Institute, the technology wing of the National Council of Education. Sarkar laments that,

The non-industrial, non-technocratic, non-engineering and non-commercial beginnings of techno-industrial education in Bengal cannot be overemphasized. The twentieth century commenced for the Bengali people as a throughly primitive and medieval age, as an anachronism of the modern world. The technoscientific backwardness and economicocultural lag in India vis-a-vis the more advanced regions of Eur-America was to be measured, so far as educational institutions are concerned, virtually by the entire period since the beginning of michanism and industrial revolution.[19]

His sentiment is echoed in the first report of the B.T.I. published in 1908:

Our attention is first directed to the great industrial awakening that has of late taken place in the country. Everywhere the signs of this new industrial spirit have manifested themselves. Our eyes have been opened to the vast natural resources of the country lying practically undeveloped. The success of Europeans in the field of industrial enterprise has also been an object lesson to us and has stimulated our energies.

Generally speaking, our country possesses no institution where higher technical instruction on modern times is imparted. Students have to go abroad to get the requisite training. But only a limited number can avail of this opportunity. Technical institutions in the country are, therefore, an absolute necessity. To meet it in some measure, however, inadequate, the society for the promotion of technical education in Bengal was organized and the Bengal Technical Institute was founded in pursuance of this object.

They took for their ideal of technical education the systems that prevail and have succeeded so well in Europe and America at it was thought advisable that so far as it lay in their power, the system should be shaped on those models. The society for the promotion of Technical Education in Bengal was thus started on the 1st June 1906 and was formally registered under Act XXI of 1860. The main object of the Society is to impart scientific and technical education to the Indian people, calculated to further their industrial progress.[20]

Sarkar describes the tug-of-war between the Bengal National College and the Bengal Technical Institute over the priority of the one over the other, Sarkar summarizes this conflict thus,

From 16th November 1905 to 25th May 1910 the people of Bengal developed a regular Kultur Kampf (Cultural War) in the educational world. One party had its centre in the National Council of Education (with Bengal National College and School), the other in the Society for the Promotion of Technical Education in Bengal (Bengal Technical Institute). The war was fought over the question of the proper attitudes of the nation towards the Government's Department of Public Instruction and the Calcutta University. The National Council wanted to function as a University and a technische Hochschule in combination, a Harvard University together with a Massachusetts Institute of technology. Its aim was (1) to grow into a full fledged omnibus and poly facultied University, capable eventually of replacing the Calcutta University in due course and (2) at the same time to promote mistrification or technocratization of all grades through schools and colleges. The society had no such ambitions but wanted simply to function as a body for the imparting of industrial education such as had been ignored by the Calcutta University and Department of Public Instruction. The Society believed that Calcutta University was minding its own business well, so far as it went, and did not care to interfere in the fields in which it had been active for nearly half a century.

The National Council's orientations to the University and Department of Public Instruction were radical and revolutionary. The shortcomings of the

University's and the Government's educational work were declared to be threefold. First, it had been systematically over literary, i.e., it had neglected exact science and technology. In this charge against the University and the Department the position of the National Council agreed to a certain extent with that of the Society, but it really went much further than the latter. The National Council was not convinced that due attention had been paid by the Government and the University to the study of the modern sciences (physics, chemistry, geology, biology, medicine, public health etc.). In the course of fifty years India had not been able to produce bands of scientific researchers. With one or two exceptions modern Indians had no address in the world to exact science. No encouragement had been forthcoming from the University and the Department to promote researches and investigations by Indian scientists and to spread the knowledge of the natural and positive sciences among the people. The Society's complacency and modernism in regard to the Department's and the Universities up-to-date accomplishments were therefore unthinkable to the National Council.

The National Council would of course initiate education in the applied sciences of all sorts. The utilization of the natural resources of the country would by all means be the most fundamental plank in its educational programme. Mechanical and electrical engineering, applied chemistry, metallurgy, mining and other branches of technical and industrial education would demand its utmost attention. But it would intensify and generalise also the cultivation of the theoretical aspect of all these physical, natural or positive sciences in a manner befitting the interests and requirements of the teeming millions. The divergence of views, between the National Council and the Society was indeed profound. The N.C.E. could not afford to remain a mere institution for the promotion of mistrification, pure and simple. Liberal arts, culture, humanism etc. were also to be promoted. Besides, its scheme of mistrification was of a higher order and more scientific or theoretical than of the Society.

In the second place, the National Council charged the Department of public instruction as well as the University with having neglected, overlooked or ignored the "national ideals". Adequate importance had not been attached to study of India's languages, literatures, history, philosophies, fine arts, as well as economic and political achievements. The University had failed to promote research into the civilization of ancient and medieval as well as modern India. Last but not least, Bengali, the mother tongue of the people, had not been made the medium of instruction in the primary, secondary, collegiate and University stages.[21]

For all practical purposes, the Society (and Bengal Technical Isntitute) began to be described as "moderate", "loy a list", progovernment and so forth while the National Council was taken to stand for "extremism", freedom, anti-Government, and what not. The dichotomy was perfect in the speculative sphere as bazar gossip.[22]

Sarkar made the following observation on this conflicting situation.

This polarity of views conveniently boiled itself down into the antithesis nationalism vs. materialism and culture vs. crude mistrification. As the exponent exclusively of technical education the society was condemned by certain sections as materialistic, unpatriotic, denationalized, antinational or Western minded. On the other hand, national glory, patriotic idealism, historic culture-sense, allround nationalising mistrification combined with science and culture and politics, all then virtues, were attributed to those who worked for literary, scientific as well as technical education such as might be a rival to and ultimately supplant the existing system administered by the Government Department and the officialized University. Much of this antithesis, dichotomy or polarism was indeed verbal. But the slogan, Bengal National College vs. Bengal Technical Institute, served to furnish the raison d'etre of the tug of war in the nationalist world of Bengal for over four years from November 1905 to May 1910.[23]

In spite of its thorough-going bread-and butter philosophy and pure materialism or crude mistrification, Bengal Technical Institute failed to attract an adequate number of scholars for industrial education. Nor were the patriotism, national idealism, swadeshi-swaraj philosophy and mistrification-cum-culture and science ideology of the Bengali revolution effective urges for education, literary, scientific as well as technical. The status quo embodied in the department of public instruction and the university was too powerful. The bastille of medievalism, anti nationalism, non-scientific and non-techical literarism could be subverted. And so the tug of war between the two wings of nationalists, moderate and extremists, came by degrees to a close. Each had found its own measure as well as that of the other.[24]

The Kultur Kampt was ended by the dissolution of the society and the total merging of that body into the National Council in May 25, 1910.

With the cessation of hostilities the National Council scored a formal triumph over its rival, the society, in regard to the question of ideals. But it was a glorious hara-kiri that the Society committed. For in the course of a few years, be it said at once, Bengal National College found itself with empty benches and automaticaly ceased to exist. By 1916-17 not a student cared to come for literary and scientific instruction along national lines. The institution that endure and survived is the institution of alleged materialism of bread-and-butter (dal-bhat) philosophy, Bengal Technical Institute. The Society for the promotion of technical education was therefore justified by nemesis. In the interest of intellectual honesty, this justification of the society has need to be frankly admitted today by all those who, like the present author, were in the opposite camp, i.e., exponents of N.C.E. ideology during the epoch of the Bengali revolution 1905-10.

This sums up Binoy Sarkar's views on Technical Education, Sarkar was evidently in favour of swadeshi, industrialization through an amalgamation of both nationalism and materialism, culture and crude mistrification which he visualised for the National Council of Education. He was however in favour of a proper college of Engineering and Technology which would upgrade the level of mistrification into technology for nation-building purpose.[26] He points out that,

> The teaching of mechanical, electrical and chemical engineering were not undertaken by the Calcutta University though Asutosh Mukherjee managed to absorb much of what N.C.E. was doing on nationalist lines, as its vice chancellor. The curriculum of the proposed college of Engineering and Technology, which grew out of the B.T.I. since 1917 and became functional from 1928. It was popular with the student community of Bengal and filled up the gap left by Calcutta University.[27]

This was a wish fulfilment for Binoy Kumar Sarkar, the prophet of swadeshi industrialization. He never tired of the glorious achievement of the Jadavpurians in the field of engineering and technology. In his reminiscences, he proudly narrates the temper of Jadavpur, its ideology, its contribution to trade and industry. He counts at least five hundred swadeshi technologists like Alamohon Das in 1946 who were only two hundred in 1925 and hardly 100 at the beginning of 1905. They mostly came from the Jadavpur campus; wherever in India there were industries involving instrumental and chemical engineering, men from Jadavpur could be found in harness. Besides, the technocrats of Jadavpur had set up many swadeshi industries themselves. They were the pivots of this swadeshi industrialization and their contribution to Bengali industrial culture has made them immortal.[28]

References

1. Concept of National Education in India, Diamond Jubilee Seminar of WCE, Calcutta 1967, p. 98.
2. Haridas Mukherjee (ed.), Binoy Sarkarer Baithake, Calcutta, 1942, p. 321.
3. Haridas Mukhopadhyay, Dawn Societer Satis Chandra Mukhopadhayay (in Bengali), Calcutta, 1948, passim.
4. The Dawn Society and the Dawn Societies Magazine, 1897-1913, Calcutta, see relevant volumes.
5. Haridas Mukherjee, Binoy Sarkarer Baithake, *op.cit.*, Introduction.
6. Binoy Sarkar, Education for Industrialization, Calcutta, 1946, passim.
7. C. Palit & S. Pahari, Satis Chandra Mukherjee the Dawn Society and National Science, Kolkata, 2002, pp. 1-6.
8. C. Palit, Growth of Commerce and Industries in Bengal, Kolkata, 1999, pp. 12-13.
9. Binoy Kumar Sarkar, Education for Industrialization, Calcutta, 1946, p. vii.
10. *Ibid.*, p. 3.
11. *Ibid.*
12. *Ibid.*, p. 5.
13. *Ibid.*, p. 17.
14. *Ibid.*, p. 18.
15. *Ibid.*, pp. 21-22.
16. *Ibid.*, p. 22.

17. *Ibid.*, pp. 77-78.

18. *Ibid.*, p. 78.

19. *Ibid.*, p. 88.

20. *Ibid.*, p. 91.

21. *Ibid.*, p. 96.

22. *Ibid.*, p. 97.

23. *Ibid.*, p. 97.

24. *Ibid.*, p. 98.

25. *Ibid.*, p. 94.

26. *Ibid.*, pp. 97-98.

27. *Ibid.*, p. 99.

28. Binoy Sarkarer Baithake *op.cit.*, Conclusion.

SECTION-III : MEDICINE

8

Hospital and Medical Education in Calcutta

The Beginning

Captain Alexander Hamilton who visited Calcutta into 1708 wrote; 'One year I was there and there were reckoned in August about twelve hundred English, some military, some servants of the Company, some private merchants residing in the town and some seamen belonging to the shipping lying at the town and before the beginning of January, there were four hundred and sixty burials registered in the Clerk's book of morality'. Naturally, Hamilton concluded that Job Charnock could not have chosen a more unhealthy place on all the river.[1]

Quoting the above, Dr. J.R. Martin, surgeon to the Native Hospital, continues in his note on the medical topography of Calcutta and its suburbs in 1834; 'In most recent times, it was the custom of the European inhabitants of Calcutta to meet on the 15th of December of each year to congratulate each other on their escape from the period so emphatically marked by Captain Hamilton.[2]

Hamilton traced the cause to the marsh and rank vegetation producing constant unwholesome exhaltions retained and accumulated in a humid atmosphere. Dr. Martin who visited the native quarters regularly also blamed the rankest compound of villainous smells that ever offended nostril and caused remittant and intermittent fever with all their fatal consequences which prevail to an extraordinary degree in all parts of the city and during every month of the year .... spleen, dropsy and bowel complaint ...... arise from marsh and other exhaltions.[3]

Martin was surprised to find that unlike in Ara, his previous posting where almost all natives slept on raised floor, in Calcutta ninth-tenths of the entire population slept on the ground. The congested mass of buildings without opening, the want of facilities for removing filth in an area like *Burrabazar* could flare up contagious diseases into an epidemic.

In suburbs like *Kurbalah, Simlca, Mirzapore, Pasitollah and Collingah,* defective drainage and half-dried tanks abounded and the drains of Ballygunge were all in a neglected condition and required cleaning with a proper fall. There was much bamboo and other jungles all around to be rooted out. He was frank enough to admit that 'we are far indeed from having effected for our emporium' all that we ought to have done for it. For many years to come, little more can be expected from that commercial and private enterprise to which Calcutta has hitherto owed so much.

The disease that bred in the native quarters could create havoc in the white town. Yet the enlightened and liberal government had done precious little to launch a general hospital for the Calcutta public. All efforts towards that had been only for the white population-sailor soldier, clerk, merchant or the officialdom.

It was the common practice since the inception of the East India Campany in 1600 to have British doctors on board their ships to the East. But when Calcutta had too many white settlers, sailors and soldiers, a hospital was contemplated. The earliest was in 1707 in the Garstin Place area, with the Tank Square on one side and the burial ground on the other. High mortality rate explains the co-existence of disease and death; a hospital from which few returned, says Hamilton in 1708.[5]

The burial ground garners the last remains of Job Charnock at the feet of St. John's Church erected in 1716.

This first-ever fever hospital was razed to the ground in 1756 at the time of Nawab Sirajudaulah's siege of Calcutta. A provisional thatched hospital was raised in the precincts of Fort William which ran about 14 years. In 1768, the Govt. purchased the huge estate of John

Zachariah Kiernander for one lakh rupees. The Reverend himself undertook construction of some additional blocks. The complex was completed in 1770. The Presidency General Hospital, as it was than called, was meant only for the Europeans.[6]

But still there was no native hospital for people at large. The first proposal for a native hospital was mooted by Rev. John Owen, a chaplain at whose suggestion a public meeting was called at La Gallia's Tavern on 27th September 1792. A sum of Rs. 12,000 was raised on the spot and the government agreed to provide Surgeons from among the Company's servants to start with. In the following years, the subscriptions amounted to 54,000 rupees. Of this, Lord Cornwallis gave Rs. 3,000, each member of council Rs. 4,500 and the Nawab Wazier Rs. 3,000. The government further consented to supplement it by a monthly subscription of 600 rupees and free supply of medicines from the Company's dispensary. The hospital was opened to the patients on 1st September 1794 by Sir John Shore in a hired house. It was later moved to its new premises in more airy and open area of *Dhurrumtala* which had only 3 or 4 houses at that time. An account of 1798 gives a good idea of the hospital's activities. It had 209 house-patients, 464 out-patients of which 523 had been relieved and discharged, 36 died and 57 was still under treatment. It had two branch dispensaries, one at *Gariahatta* and the other at *Kalinga*. The native hospital remained in this site for long 75 years. As Dhurrumtala became crowded, it was proposed in 1871 to remove the hospital to the more salibrious climate on the river bank near the mint. The old establishments were sold for Rs. 97,000, public subscriptions fetched Rs. 78,900 and the Mayo Memorial Fund donated Rs. 50,000. On 3rd February 1873 Viceroy Lord Northbrook laid the foundation stone of the proposed hospital and it became functional from 5th September, 1874. It was renamed Mayo Hospital.[7]

The first Maternity Hospital known as the Calcutta Lying-in Hospital was established on 12th June 1814 on Park Street. This was due to the joint efforts of the wives of the Count Loudon and Moira.[8] But a fully equipped hospital for the natives was still a far cry. A Lottery Committee was set up in 1817 to raise money for the improvement of Calcutta. The money raised from the public in this way was spent mostly for mansions and carriageways in the white town and there was little of it was actually diverted to the native quarters. The government had never given any attention to and spent money on such improvement of the city. In 1833, this dubious way of raising money for municipal works was given up due to the pressure of public opinion in England. This was replaced by a Fever Hospital Committee by Lord Auckland in 1835.[9] The prime mover in this regard was Dr. James Ronald Martin, Surgeon to the Native Hospital. He addressed a letter to the governors of that Institution pointing out the urgent necessity for establishing a Fever Hospital in a central part of the native town to afford medical relief to the poor natives among whom there was almost a frightful prevalence of fever and other diseases. The note on the medical topography, he subjoined with this appear, has been already quoted at the outset. Martin who actually visited the dingy blind lanes of the black town suggested that the masses of buildings should be demolished to make room for planned housing. He emphasized the need for airy and sunny streets, required pavements, proper scavenger service and construction of a scientific network of sewers connected with sewer from the houses, supply of pure drinking water at convenient places, filling up of pits and half-dried ponds, etc.[10] Martin knew that diseases for the native quarters could flare up into epidemics affecting the white establishments. There should not be any dichotomy in public health policy. Much of the municipal funds had also been raised by lotteries and subscriptions from the natives. It was, therefore, meet and proper that a Fever Hospital should be launched without delay for all citizens of Calcutta. In spite of such a fervent appeal, a sub-committee was formed which had Sir E. Rayan, Sir J.P. Grant, Sir J.R. Martin and eminent Indians like Ram Kamal Sen, Ram Chandra Das and Radhakana Deb. Funds were contributed by Europeans and Indians alike. It amounted to Rs. 47,713-7 as 5 pies. The construction was started immediately and by 1st april 1838 and 20 bedded hospital was started. In 1840, the Maternity Ward of 100 beds was added. The foundation stone of the massive complex visible from College Street was laid on 3rd September, 1848 by Lord Dalhousie and it was completed in 1852 on grounds donated by the noted business magnate Babu Motilal Seal. The final expenses came to Rs. $2^1/_2$ lakhs and the buildings stretched from the present crossing of College Street and Colootola Street to Central Avenue. The Medical College started in 1835 was also housed within this campus.[11]

As medical facilities were extended to the general public, the limited number of Company's surgeons and private practitioners could not cope with it. The need for training the native youth in western medicine and surgery was keenly felt. The first native medical institution was open in Calcutta in 1824. The medium of instruction was Hindusthani and books in that language were being written by Orientalists. Dr. John Tytler was its Principal. The next step was to open medical classes in Arabic, Sanskrit and Hindusthani in the Sanskrit College. This was also known as the medical school. In August 1831, a small clinic was attached to it through the instrumentality of Babu Ramkamal Sen and Dr. John Grant began teaching medicine in this school. The pioneer Indian medico Pandit Madhusudan Gupta began his medical career here as an apothecary and later a teacher of repute.

1833 was a major landmark for medical education in India. Lord Bentick decided to set up a committee of 5 members in that year to report on the state of medical education and suggest its improvement. The only Bengalee on the committee was Babu Ramkamal Sen. The report was submitted on 20th October 1834. On the basis of the report Lord Bentick published its circular proposing the establishment of the Medical College in supersession of the existing institution with English as the medium of instruction. The College was to be opened to all. D. Mauntford Joseph Bramby was the first Superintendent (later Principal) of the College, on 1st March 1835. Dr. Henry Harry Goodeve, the Professor of Surgery and Dissection section joined on 9th February and on 17th March, Pandit Madhusudan Gupta of Sanskrit College joined as Assistant to Dr. Goodeve. By the Government Order of 28th January 1835, an admission test was taken of willing candidates, on 1st May. David Hare and Rev. Alexander Duff persuaded students of their schools to sit for the test. About a hundred appeared out of which 67 students were admitted, 49 being stipendiary students on 1st June. The students with the exception of one Christian were all of the Hindu persuasion and of various castes, belonging chiefly to the middle classes of native society. David Hare was appointed Secretary to the College in 1837 on the death of Principal Bramley on 19th January by the newly constituted Council. It was his fervent appeal that shortly led to the foundation of the Fever Hospital attached to the College for practical work. Dr. Nathaniel Wallich, Professor of Botany and Dr. William Brook O'Shaughnessy, Professor of Chemistry and Medicine joined the staff soon after. David Hare was succeeded in 1841 by Dr. F.J. Mouat whose indefatiguable labour led to the all round development of the College.[12] The first degree examination (Lincense of Medicine and surgery) was held on 30th October, 1838 for seven days. Eleven students appeared after a session of $3^1/_2$ years. The first examination for letters testimonial qualifying candidates for the posts of sub-assistant surgeons was held on 18th December 1840. Madhusudan Gupta, Jadabchandra Dharra, Chaman Lal, Rajkisto Chatterjee, Gobinda Chandra Gupta, C. Finlay and Navakrishna Gupta qualified. Madhusudan and Navakrishna retained their situation in the Medical College and the five other young men were reported to the Medical Board and to the Government as being available for the public service as sub-assistant surgeons.[13] In 1839 and 1852, the Hindusthani and the Bengali departments were specially reorganised to recruit and train native doctors for the military establishments in the upcountry and the interior of Bengal proper for their acceptability to the native population. Medical and scientific books were suitably translated.[14]

A return of the sub-assistant surgeons educated at the Medical College of 1848 showed 53 of them posted in various parts of India. Because of low pay, the College failed to attract students. The pay of the sub-assistant surgeon was revised in 1848 as follows:

7 years service : Rs. 100 p.m. : 7-years and under 14: Rs. 150.00 :14 and upwards : Rs. 200.00 15. Madhusudan Gupta after a long career of 13 years qualified in 1848 for a salary of Rs. 200.00. The colonial authorities employed this cheap yet meritorious agency of native doctors where white doctors at high pay were not

available. Their main purpose was to find native assistants to white surgeons in Calcutta and the military and civil stations in the sprawling hinterland to insure health for such establishments in the interior. It was evidently public health policy in name and not strictly meant for the Indian public in general. But disease was no respecter of colour or creed. Barefoot native doctors were really needed for mass communication and control of epidemics. The enclaved nature of the health service circumscribed its growth and potential for health insurance.

Despite these limitations, certain milestones in the history of the Medical College may be noticed. The most momentous of them happened on 28th October 1836 when Dr. Goodeve led the first team of Indian apprentices to the work of dissection of the dead body to teach practical anatomy. Dr. Bramley, Principal of the College reports:

Dissection is seldom approached by the non-initiated even in Europe without feelings of aversion and it was much to be readed lest the first impressions upon these youths who were peculiarly sensitive to every feeling of the kind should operate to alarm and discourage them from the pursuit. It was, moreover, necessary to conduct the dissection with due regard to secrecy as the students were naturally enough exceedingly averse to being exposed to the gaze of the intruders. Particularly as such exposure might entail the penalty of excommunication to themselves and their families .... and I have the gratification to say that on the 28th October, all doubt was removed. On that day which may be regarded as an eventful era in the annals of the Medical College, four of the most intelligent and respectable pupils at their own solicitation undertook the dissection of the human subject. The hero of this silent revolution was Madhusudan Gupta who took the knife first but Umacharan Seth, Rajkrishna De, Dwarkanath Gupta and Nabinchandra Mitra also performed the operation after him. This deed was a red-letter day in the history of our renaissance when science triumphed over superstition in a dramatic way.[16]

To encourage the students, Prince Dwarkanath Tagore instituted prizes for excellence in various branches of the study of three years at Rs. 1,000 per annum for three years, in March 1836. Other donors included Ramgopal Ghosh, Rustomji Nawasji and Ramkamal Sen. Madhusudan Gupta was presented a self-potrait with the words, 'the first Hindu Anatomist of British India' written into it by Dr. Goodeve and Bethune.[17]

The second memorable milestone was the princely way in which Dwarkanath Tagore sponsored three medical students to go to England for higher medical studies in 1804. The fourth scholarship was offered jointly by the Nawab 'Wazier of Bengal' and Dr. Goodeve of the College who escorted the lucky four to England. They were Gopal Chandra Seal, Bholanath Bose, Dwarkanath Bose and Surya Coomar Chakraborty who duly qualified for their MRCP and MRCS degrees in 1846 and Bholanath went on become the first M.D. in 1847 standing second in first class. In 1855, Indian Medical Service (IMS) examination was conducted for the first time and Surya Coomar proceeded to England for a second time to sit for the examination. He stood first and was recruited to the convenanted service at par with the white candidates. He came back to join as the assistant professor of medicine in the Medical College. He rose to become its senior professor. He was the first Indian to enter the European preserve.[18]

The third landmark was the development of the Medical College as the best and the most coveted institute of scientific and medical research. The personal anecdotes of Dr. Mahendralal Sircar illustrate it very well. Mahendralal joined Hindu College in 1849 to study pure science and mathematics. He was extremely proficient in mathematics and very soon became the dearest student of the Principal and other teachers of the College. But it was the fashion of the aristocratic and wealthy families to put their wards in the liberal arts and preferably in English literature. Science was neglected as a field of study. The teaching of science was evidently unsatisfactory as the nickname 'Soda' was ascribed to the teacher of Chemistry[19]. Mahendralal was avid for the pursuit of science and in 1855, he took admission to the Medical College where science was most seriously taught by eminent teachers like Wallich and O'Shaughnessy, Goodeve, and others. Scientific education was crucial for public health and therefore not peripheral here. Mahendralal stuck to his decision despite persuasions by his teachers and relatives. He got here what he wanted. There was no turning back and he dedicated himself to his new field of study. He surprised his teacher of Optics, Dr. Archer one day by answering critical questions on optics which the learned professor was teaching to the 4th year class when he himself was a sophomore listening to the lecture from a distance. Dr. archer became so pleased that he wanted him to deliver a lecture on Optics to the senior students.[20]

The Medical College became the centre of higher studies and research due to the research and teaching undertaken by its galaxy of eminent teachers. These talented men failed faced with the tropical diseases began new research into this untrodden area. The result was a remarkable advance in the medical science for the world at large. The case of Malaria can be cited. It was a very common and virulent fever which periodically enervated the native population. Its cause was unknown and so also the correct antidote. The cause was ascribed to damp climate and exhaltion from the marshes. The use of cinchona for Malaria was well known. In 1638, the wife of the Spanish Viceroy to Peru. Countess Anna del Chinchon was down with this fever and a friend of the Viceroy sent the barks of a plant for cure which really had the healing power. This led to the processing of quinine from Cinchona, a name derived from her title as a medicine for Malaria. But the cause of the fever was unknown.

The bacteria was discovered by Dr. Laveran in 1880 while working in Algiers. But the source was still a mystery.[20] After almost sixty years of the foundation of the Medical College and ninety years after the establishment of the Presidency General Hospital, Ronald Ross working in the laboratories of the latter finally proved that the *Anophales* female mosquito was carrier of the germ and injected it into the bloodstream of her human victim. Meanwhile Malaria had become the scourge of the countryside in Bengal and other provinces and all species of tadpoles and insects were suspected and experimented upon till on 20th April, 1897. Ross discovered the germ in the stomach of the mosquito in his Calcutta laboratory. Ross was born in Almora and spent his life in Secunderabad and Calcutta, an Indian by domicile.

He won the Nobel Prize for this epoch-making research.[22] Some others like Upendranath Brahmachari, the explorer of *Kala Azar* would form the select band of Indian scientists who made medical history in the same way. The School of Tropical Medicine which came into being in the 1920s was the crowing glory of the East-West collaboration in medical research and treatment of undeciphered diseases.

References

1. Note on the Medical Topography of Calcutta and the Suburbs chiefly with reference to the condition of the native health. Appendix A to the Report of the Committee

appointed by the Rt. Hon'ble the Governor of Bengal for the establishment of a fever hospital etc. 7 January 1810, p. 153 onwards of the Guide to the Records, Part 1, 1759-1858.

2. *Ibid.*
3. *Ibid.*
4. *Ibid.*
5. Captain Hamilton, A. (1708), New Account of the East India Company, London, quoted in Chakrabarty, A. (1975) Chikitsa Vignane Bengali (vernacular), Calcutta, pp. 21-22.
6. *Ibid.*, p. 23.
7. *Ibid.*, p. 24; also Banerjee B.N., Sambadpatre Sekaler katha (Bengali) Vol. 1, 1818-30, pp. 205-16.
8. Bagal, J.C. (1959) Kalikatar Sanskriti Kendra (Bengali) Calcutta, pp. 83-90.
9. *Ibid.*
10. Infra 1 : Report of Dr. Martin. J.R. dt. 9 April 1835.
11. Bagal. J.C. *op.cit.*
12. Report of the General Committee of Public Instruction (GCPI), 1836-44, quoted in Palit, C. (1980), New Viewpoints on Nineteenth Century Bengal, Calcutta, pp. 68-69.
13. *Ibid.*
14. Bagal, J.C. *op.cit.*
15. Palit, C., *op.cit.*
16. *Ibid.*
17. *Ibid.*, and Bagal, J.C. *op.cit.*
18. Chakrabarty, A. *op.cit.*, pp. 67-77.

19. A Century, Centenary Volume of the Indian Association for the Cultivation of Science, Calcutta, 1976, pp. 4-24.

    Also, Sivnath Sastri, Ramtanu Lahiri O tatkalin Bangasamaj (Bengali), 1909 New Age Edition : See Mahendralal; also C. Palit, Mahendralal Sarkar, 1833-1904 : The Quest for National Science in Indian Historical Review, 12 nos. 1-2, pp. 121-128.

20. Chakrabarty, A. *op.cit.*, p. 47.

21. *Ibid.*, p. 44.

22. Bentley, C.A. (1916) Malaria in Bengal Part I, Calcutta Chapter 1.

9

Ayurveda in Colonial Bengal : Survival and Revival

Precolonial History

Ayurveda or the science of life was not the monopoly of the *Brahmins*. It was open to all four castes who could pursue it for religious merit, wealth and pleasure, according to *Charaka Samhita*[1]. The healing arts formed part of secular knowledge and had lay practitioners, the *visaj*, the *ambasthas* and the *chikitsakas*. It was a need-based, utilitarian science but came to be codified in theory and practice in the texts of Charaka, Susruta and Bagbhata by 5th century A.D. Though hostilities between priest and lay physicians are reported in *Manu Samhita* and other *Brahmasutras*, the thoery of *Brahminical* spiritualisation and domination of Ayurveda and its consequent decline as a secular science as if it were primarily developed by non-*Brahmins* is unhistorical [2]. There was, in fact, considerable accomodation of secular knowledge like Ayurveda within the sacred texts and it was developed by both *Brahmin* and non-*Brahmin* scholars from *Atreya* to *Bagbhata*. The conflict was more between practitioners of medicine and of surgery as *Brahmins* concentrated more on the former and the non-*Brahmins* on the latter. Ayurveda was treated as an *Upaveda* and its absorption in cosmology was a common trend prevailing over medicine in preindustrial Europe. There was no golden past from which there was a steep decline due to Greek-Scythian or Muhammedan invasions as many revivalists tend to uphold. Considering worldwide state of the art, Ayurveda would compare favourably with Egyptian, Greek and Chinese medicine. But it had certain inherent weakness. Its physical theories of five elements (*Panchabhutas*), three humors (*Tridosha*) and seven substances (Saptadhatus) cannot be rationally reconciled to each other. Its anatomy was incomplete. In *Charaka* and *Susruta*, properties of head and heart are not properly delineated as they were not subjected to dissection. There was no clear conception whether blood was a humor or a substance. There was no knowledge of the viscera. *Susruta* is better on bones, nerve-junctures, entrails and ducts. All diseases were thought to be due to imbalance of the humors and restoration of their equilibrium was thought to be the cure [3]. There is no place for germ theory or genetic decay of organs in Ayurveda. There is over reliance on texts or *Aptavakyas* and neglect of experimental verification. All natural sciences had an evolution based on experience. Even perception of truth is relative and not immutable. Ayurveda became stagnant because of its textual orthodoxy uninformed by case-histories. In the days of difficult transport and communication, knowledge could not become public and universal. There were no printed books and manuscripts were hoarded as secret knowledge and source of wealth. A few fortunate possessors of Ayurvedic texts would set up their *gurukuls* which kept them confidential. *Gurukuls* and later *Tols* depended on a variety of versions of the *Bridhatroyi*, i.e., *Charaka*, *Susruta* and *Bagbhata*. Rarely induced secrecy. Its commercial value was duly exploited. The *kaviraj* or *vaidya* prepared his own medicine. As a practitioner, it was the secret of his material success. The materia medica was not disclosed or standardised. It remained crude and improvised. In Ayurveda, each individual was considered unique and the treatment was person specific. It was unable to cope with epidemics and unknown diseases caused by bacteria. Common factors and therapies were not developed for general benefit[4]. These inherent defects are glossed over by the revivalists and the blame for decline of Ayurveda is solely put on Muhammedan and British patronage of their Yunani and western medicine. The Muslim, especially, the Mughal rulers were not at all inimical to Ayurveda. Though they brought the Yunani system of medicine with them, they also encouraged Ayurveda as a cure. Harun-al-Rashid is known to have welcome a vaidya to his court at Baghdad for treatment. As Basham observes; 'The practitioners of the two systems seem to have collaborated because each had much to learn from the other and whatever the *ulama* and the *Brahmans* might say, we have no record of animosity between Hindus and Muslims in the field of medicine'[5]. Poonam Bala has given the whole list of integrated medical works and institutions partronised by the Mughals since Babur [6a,b,c]. Ayurvedists also absorbed considerable knowledge from Yunani medicine. Blood as an additional humor or substance, sphygmology or pulse lore, use of mercury and opium in specific cases were concepts directly derived from the Yunani system[7]. There is, therefore, little basis for the revivalist propaganda that Ayurveda declined due to Muhammedan rule. In fact, the integrated approach made it a living science. Before concluding this section, it can be said that though in Europe, a classical ideal was used to oppose medieval scholasticism during renaissance, scientific knowledge was demythologised with the scientific revolutions in the following centuries and the literal authority of the classical age was challenged. In India, the revivalists clung to the literal texts of ancient times despite progress of medical science since then. Even the integrated approach prevalent in precolonial India was given up.

Colonial Status

From Warren Hastings to William Bentinck, colonial rule was precarious not only due to traditional resistance it had to smother and pacify through coercion and conciliation but also due to the rigours of the tropical climate and diseases which threatened the physical existence of the ruling class. They were avid for all kinds of information about the land and its people-its flora and fauna, mineral resources, language and literature, law and administration, science, technology and medicine, religion, manners and customs. A spirit of paternalism prevailed during this period of collection and compilation of information in order to classify and control it for effective rule. The Asiatic Society (1984), the Botanical Gardens (1987), the Serampore Mission (1799), the Fort William College (1802) and the Agrihorticultural Society (1820) were the fora through which information was gathered. William Jones, the founder of the Asiatic Society was the first to write perceptively on indigenous medicine in his 'Botanical Observations on Select Indian Plants' between 1790-1800. In 1813, the Court of Directors of the East India Company recommended to its government in India the study of Indian tracts on medicine, 'the knowledge of which might prove desirable to the European practitioners'[8]. This was followed by Whitelaw Ainslie's Materia Medica of Hindoostan. In 1822, the Native Medical Institution (NMI) was established and in 1826, parallel instructions in Ayurveda and Unani medicine was started in Sanskrit College and Calcutta Madrasa along with teaching of western medicine through translated western texts. In 1833, more teachers were appointed to teach *Charaka*, *Susruta* and *Bagbhata* in these institutions. Both NMI and Sanskrit College continued this practice till 1835. Pandit Madhusudan Gupta began his career as an Ayurvedic teacher at

Sanskrit College. Students qualifying from these institutions were absorbed as native doctors in civil and military establishments. This policy of accommodation was, however, given up in 1835 after the so called victory of the Anglicists over the Orientalists but really after the consolidation of British rule in Bengal had given the authorities the confidence to override the indigenous system and introduce western medicine. The NMI and the departments of indigenous medicine at Sanskrit College and Madrasa were abolished by Government order of 28 January 1835.

Calcutta Medical College as the pioneer of western medicine took their place with English as the medium of instruction. Madhusudan Gupta was retained on the staff of the CMC as the sole survivor[9a,b]. Madhusudan Gupta led the dissection of the corpse at the CMC against caste taboo and *Shastric* injunctions. He became the hero and harbinger of western medicine and was presented a self-portrait with the words 'the first Hindu anatomist of British India' written into it by Dr. Goodeve and Bethune. But from the Ayurvedist's point of view, it was a case of surrendered identity by the scion of an Ayurvedist family of repute. The cannonfire salute of fifty rounds from the Fort William simultaneously sounded the death-knell of Ayurveda. By adjuring Ayurveda, Madhusudan was guilty of national betrayal. He qualified as a sub-assistant surgeon for the Health Service and though versatile in both indigenous and western medicine, ended his career carrying a salary of Rs. 200 per month[10a,b]. Though supersession was a matter of policy, appropriation of indigenous medicine in British Pharmacopoeia continued unabated as they were cheaper and easily accessible tropical medicines unknown to the authorities. Thus Brooke O' Shaughnessy compiled Bengal Pharmacopoeia in 1837 dealing exclusively with the properties and uses of the medicinal plants in Bengal. A medical laboratory was established in Calcutta to process and prepare drugs from them. The indigenous materia medica was tapped and authorities consulted in selecting and standardising such drugs for test in hospitals and charitable dispensaries all over Bengal. Several such drugs like *Cannabis indica* were added to the British Pharmacopoeia. During 1839-40 cholera epidemic, country medicines such as *kaladana, but kelija* and drugs composed of opium and calomel were extensively used following their success at Chittagong dispensary[11]. In 1868, the Government published the Pharmacopoeia of India. The eventual purpose of this scientific activity was to appropriate many indigenous into British Pharmacopoeia[12]. It was the government policy to insist on low-cost indigenous drugs against the wishes of allopathic practitioners who wanted western medicine imported from England at an enormous cost. The Medical Department in 1866 ordered that indigenous drugs should be used more generally and 'the ones available in a crude and uncouth form be given laboratory treatment before administration' and supplied to all government dispensaries in the presidency[13]. Along with this appropriation of drugs, the rural kaviraja and hakeems were also systematically absorbed within the Health Service. In an official report of 1870, the following table of doctor : people ratio under the Health Service was given for some districts of Bengal:

On an average, 1 doctor of the above group was available for every 2000 people for all Bengal. This was hopelessly inadequate. Nor could poor people in the countryside afford costly European drugs like quinine. Indigenous herbal medicines were used by allopathic doctors. The colonial government was unwilling to provide the required number of staff trained in western medicine for rural Bengal. The skeletal Health Service only served the government establishments in the urban areas. So there was an effort to affiliate the *kaviraja* and *hakeems* of the mofussil in the Health Service and encourage them to adopt western system while they practised indigenous medicine. This was the least expensive way of introducing European medicine in rural area. During cholera and malaria epidemics, this was the subordinate paramedical force invariably used[14a]. But still the staff was inadequate to tackle the emergency and the rural people had to depend on their village *Kavirajas* and *dais.* We get a glimpse of the situation in Hooghly in the 1870s from the memoir of Carstairs:

Bengal's one Sanitary Commissioner used to go about making suggestions which were seldom attended to save near cantonments. Serampore's one Government

Doctor fully occupied with the mills, hospitals, patients and town work he had to attend to, seldom saw the villages, save on one of his rare and hasty visits of inspections to a country dispensary or when sent for by some rich invalid or extremist. In the towns, we had one or two qualified medical men whose hands were full of private practice. It did not pay such men to settle in the villages where it was so difficult to move about and where the fatalist Indian grudged their fees. 'Doctors'-ignorant quacks abounded in towns and villages alike. The medical attendents ordinarily employed by the people were the kaviraj-the hereditary physician who, if without a diploma inherited a good stock of experience and the 'Dai' or nurse-midwife.[14b]

Thus, the *kaviraja* and Ayurveda survived despite official patronage of western medicine. But they were increasingly subsumed under its hegemony in the late 19th century. The Government opposed all moves by the *Bhadraloks* to hoist an Ayurvedic institute of education and research. A petition by Ayurvedists for Government grants for such purposes was turned down on the ground that Ayurveda was unscientific. Even private donations to promote the cause were not utilised by the authorities between 1891-93[15].

At the same time, most of the indigenous drugs of tested value were absorbed in the western pharmacopoeia. Western medicine made giant strides towards becoming a highly organised industry by standardisation of drugs through constant experimental verification and refinement, advance in bacteriology, pharmacology, immunology and chemistry and improved techniques of industrial manufacture of drugs reducing cost. Indigenous medicine lagged far behind. The rise of the drug industry in Britain from the late 19th century and its expansion abreast with German pharmaceutical products continued unabated during the interwar period till it reached a turnover of 58 million pounds in sales in Britain by 1946. It had its inevitable impact on colonial India. Due to increased professionalisation, both British and Indian practitioners of western medicine switched over to synthetic drugs. Thus, the dominance of western medicine was virtually complete[16].

Roots of Revival

The domination generated protest in due course. Ayurveda had its roots in rural areas, in *Tol's* and *Gurukul's* left to themselves by the Raj. Allopathy was essentially an urban phenomenon confined to the metropolis and a few other towns. *Kaviraj* or *vaidya* families continued their hereditary profession in the mofussil. Brahmananda Gupta gives a list of eminent *Kaviraj* families running into three generations spearheaded by Gangadhara Ray (1780-1885), Gangaprasad Sen (1824-1896), Haranchandra Chakrabarty, Sreecharan Sen, Bijoyratna Sen, Gannath Sen, Jaminbhusan Ray, Bijoykali Bhattacharyya, Bimalananda Tarkatirtha, Prabhakar Chattopadhyay, etc.[17a,b]. Gupta also cites leading schools of traditional Ayurveda in 19th century Bengal:

School	*Speciality*
East Bengal	
Savar	Preparation of herbal medicine.
Matta	Techniques of examining patients.
Gaila	Diagnosis and prescriptions and preparation of medicines of chemotherapy.
Chandsi	Healing, many kinds of ulcers, fistula and piles.
Chittagong	Treating insanity.
Khandarapara	Treating insanity and construction of hut-like temporary hospitals.
West Bengal	
Murshidabad	Reading pulse and diagnosis.
Kumartuli	Medicine (pills of Nilambar do not fail).
Srikhanda	General physicians who sold medicines in the open market and published Ayurvedic texts.[18]

Tol education was not purely theoretical. Students accompanied the *kaviraj* whenever he visited his patients, thus acquiring clinical experience. They also had practical knowledge of collecting raw materials for medicines and preparing drugs under his direction. Gangadhar Ray, the legendary *kaviraj* of the 19th century was born in village Magura in East Bengal and studied in the Tol of *kaviraj* Ramakanta Sen of Vaidya Belghoria. He practised for a while in Calcutta in 1819 but finally moved to Saidabad in Murshidabad where he set up a Tol and trained a number of brilliant students. He acquired abiding fame in therapeutic use of poisons and decoctions, in diagnosis by reading the pulse and in prognosis. He wrote Sanskrit commentaries on 34 texts and composed 41 original works. His commentary on *Charaka* titled *Jalpakalpa* is his lasting contribution to Ayurvedic scholarship. He died in 1885 at the age of 87 holding away over Bengal Presidency even under colonial constraint. While Gangadhara carried out his crusade for Ayurveda in the countryside, Gangaprasad Sen did it in Calcutta, the citadel of western medicine. Born in Vikrampur, Dhaka in 1824, he learnt Ayurveda from his illustrious father, Nilambar Sen. He began practising in Calcutta under the shadow of Calcutta Medical College. He began preparing and selling medicine as a reply to import of western drugs with a fixed price list and due publicity in the newspapers. As a practitioner, he charged fees which often surpassed those of the British physicians. He published the first Ayurvedic magazine in Bengali. *Ayurveda Sanjivani* as a campaign for the Indian system. He also opened a Tol at home and offered free room and board to his students. He had many eminent patients including Ramakrishna. He died in 1896 as one of the richest men in Calcutta. Gangadhara Ray and Gangaprasad Sen along with their numerous students authored the Ayurvedic renaissance around 1870. It has to be remembered that in the Bengal Census of 1872, only 3769 physicians, surgeons and doctors of western medicine recorded against 23,700 vaids and hakeems. Bijoyratna Sen, a disciple of Gangaprasad Sen translated the *Astangahridaya* of *Bagbhata*, one of the *bridhatroyis* (old texts) of Ayurveda into Bengali and also published the original Sanskrit text with his emendations. He had many Europeans as his patients and his accomplishments were recognised by the Government of India by conferring on him the honorary title of *Mahamohopadhyay*. He had a good command of English and Allopathy. He sought to integrate the two systems by utilising allopathic medicines as preparations of some of his special medicines. He inspired his student, Jaminibhusan Ray to set up an Ayurvedic college where all eight branches of Ayurveda would be taught. Jaminibhusan was also a scholar of both systems. A man of the 20th century, he sought to rejuvenate Ayurveda against heavy odds. He revived all eight branches of Ayurveda by publishing suitable texts. He introduced *misra-Ayurveda* or integrated medical course in his college by adopting 65% Allopathy and 35% Ayurveda in the mixed course. This was done out of conviction and necessity to get Government grants. Kaviraj Gananath Sen, a coadjutor of Jaminibhusan in founding the Astanga Ayurvedic College was another exponent of both systems. He tried the same integration by writing a Sanskrit treatise on western anatomy titled '*Pratyaksha Shariram*' for students of Ayurveda. He was criticised by the purists or *Shuddhists* like Haranchandra Chakrabarty for considering Susruta inadequate in anatomy[19a,b].

Revivalist Strategies

What Jaminibhusan and Gananath tried was one of the ways of revitalising Ayurveda. But Purists attacked it for being neither here nor there. Those who passed the integrated course with a few papers on each system did not learn either Ayurveda or Allopathy fully. They failed to earn people's esteem as reliable doctors. What the purists upheld was the infallibility of the ancient Ayurvedic texts of *Charaka, Susruta* and *Bagbhata* which needed painstaking revision in the light of modern science. As has been pointed out at the outset its anatomy, humoral theory of disease, crude drugs and their toxicity had all to be updated and standardised. When the colonial government tightened up medical education in favour of western medicine by derecognition of practitioners of integrated medicine, stoppage of grants-in-aid and refusal to recruit Ayurvedic graduates on par with the medical graduate in the Health Service and pushed synthetic western medicine through government network of hospitals and dispensaries, the indigenous camp was divided between the integrationists and purists. Public opinion was already regimented by the hegemonic control of British and Indian practitioners of western medicine. It was a fact that the Ayurvedic colleges failed to attract the best students to their discipline and got the failed candidates of the other stream who could not claim the same status with the allopathic brotherhood in terms of pay and position. It was futile to combine both streams by learning each separately for several years. The government disallowed a condensed allopathic course for Ayurvedic students after their graduation in the main stream. The revivalists replied to these predicaments in several ways. They sought legitimacy by political means. They formed a pressure group in the Indian National Congress which passed numerous resolutions in the early decades of the 20th century for government patronage and protection of Ayurveda. The All India Ayurvedic Congress was founded in 1907 to mobilise popular support and pressurise government. The struggle for Ayurveda came to be identified as freedom struggle[20a,b]. Jaminibhusan's Astanga Ayurvedic College and Hospitals was founded in 1916. In 1921, *Gaudiya Sarva Vidyayatana*, a national university was established of which *Vaidya Sastra Pitha* was an integral part teaching pure Ayurveda. Kaviraj Shyamdas Vachaspati, the founder-principal received a donation of Rs. 6000 from Tilak Swaraj Fund. It continues even today. The Govindasundari Ayurvedic College was founded in 1922 by Kaviraj Ramchandra Mullick and patronised by Manindrachandra Nandy, Maharaja of Kashimbazar. It followed integrated medicine. The Viswanath Ayurveda Mahavidyalaya also founded in 1922 by Gananath Sen was another similar institution studying Ayurveda and Allopathy side by side[21]. The parallel education programme was vigorously maintained from private resources. With Montford Reforms of 1919 and the Governments were formed by Indian leaders and Ayurvedic medicine got some official patronage. State Faculties of Ayurvedic Medicine were formed to promote it. But all government committees to review the state of Ayurvedic education were dominated by allopathic practitioners who were generally cold towards it in recomending official support of any kind and remained undecided in commitment between integrated and pure streams[22]. The only strategy that succeeded was the manufacture of Ayurvedic drugs on a commercial scale by a few entrepreneurial Ayurvedists and chemists. Gupta writes; 'Gangadhara Ray is said to have laughed at practitioners who prepared medicines before hand calling them *Badial*, possessors of pills ready for common use. But the large scale of production of medicine became necessary as the popularity of Ayurveda grew and small scale preparations could not keep pace with the demand.'[23]. Chandrakishore Sen, a *kaviraj* of the Srikhanda School was the pioneer in the field of commercial manufacture of Ayurvedic drugs. He began it in 1878 at Kalighat on a small scale. Encouraged by success, he shifted to Kolutola to set up his factory for large scale production. C.K. Sen and Co. became a major concern and played a pivotal role in the development of Ayurveda. It also published inexpensive books on Ayurveda to propagate knowledge. A major swadeshi enterprise of the 20th century was the Ayurveda Mahamandal Co. Ltd. which was launched on 21 July 1919 at 19, Grey Street, Calcutta with an authorised capital of Rs. 50,000. It had the active backing of *zamindars*, bankers and merchants. The managing agency was entrusted to the hands of Kaviraj Surendranath Dasgupta and Jogendrachandra Sastri. Its *Viswanath Toila*, a painkiller and *Sonitamrita*, a blood-purifier were prepared through indigenous methods. In 1914,

Gananath Sen, the leading Ayurvedist and the spearhead of Ayurvedic movement in India as the secretary of Ayurveda Sabha and President of the Alahabad Session of the All India Ayurvedic Congress in 1911, launched his Kalpataru Ayurvedic Works, capable of producing drugs on a mass scale in 1914. He purchased a huge disintegrator driven by electricity which manufactured medicines on a large scale and shaped them into tablets. *Chavanprash* and *Makaradhwaja* were some of its best selling products at a cheap rate. But the crowning glory came with the establishment of Sakti and Sadhna Ausadhalayas, both having headquarters at Dhaka which exist even today. Sakti Ausadhalaya was established in 1901 by Mathuramohan Chakrabarty, a chemistry teacher of a high school, out of his salary savings. It had branches at Banaras, Calcutta and Rangpur and produced drugs on a mass scale from both vegetables and minerals. A team of *Kavirajs* were employed for the purpose. *Chavanprash*, *Makaradhwaj* and *Saribadi Salsa* were its well known products. Jogeshchandra Ghosh, the founder of Sadhana Ausadhalaya in 1914 was a lecturer in chemistry at Bhagalpur College and a student of Sir P.C. Ray. Its *Mritasanjivani Sudha* and *Mahadraksharishta* became popular recuperative tonics. It has remained in business for the last 77 years and has branches all over India and Bangladesh. Kaviraj Naresh Ghosh, son of Jogeshchandra who runs the business is also a major Ayurvedist missionary in his own right. He has presided over many provincial Ayurvedic conferences and pointed out the efficacy of cheap Ayurvedic treatment and medicine for poor masses of India who cannot afford costly allopathic medicine[24].

Postcolonial Prospect

In the freedom struggle in the field of medicine, Ayurveda had successfully countered the 'para medical subordination to the cosmopolitan medical profession' by professionalisation and institutionalisation, politicisation for legitimacy and technologically advanced manufacture of mass medicines. It was a case of modernity of tradition. As Brass writes, 'the penetration of modern medicine into the country side has been so slow that there is no real prospect of providing relief to the rural areas of India .... in the generations to come'[25]. It is in this gap that professionalised Ayurveda has come to stay. But it should not mean that the revivalist movement should score a point by asserting that all modern medical developments were anticipated in ancient India. The demand to declare Ayurveda as the national system of medicine and the cry of quit India against western medicine is sheer chauvinism. One may not superscribe to Brass's view that 'dual modernisation in India is a wasteful process' and an integrative approach should be taken to combine the best elements of both systems[26]. This would result in the assimilation of the one by the other as happened in the colonial past. The government and the vocal public are already biased in favour of western medicine. Leslie more perceptively writes: 'we should not assume as laymen, and physicians in the United States usually do, that the ideal of a uniform medical system controlled by physicians of cosmopolitan medicine is intrinsically superior to other ideals and should be the goal of all societies. The pluralistic structure of the Indian medical system might best develop toward goals that would enhance the advantages of pluralism while correcting its disadvantages'[27].

While its anatomy and humoral theory of disease are inadequate in the light of modern science, Ayurveda still retains its efficacy in chronic elemental diseases its treatment of the patient rather than the impersonal disease so relevant today in psychosomatic syndromes, its prognosis and diagnosis, pulse lore are areas deserving greater attention and improvement through logistic support of modern science. Though it is not true that all Ayurvedic drugs are non-toxic (crude, narcotic and metallic drugs do have high toxicity needing control), they are by and large more natural absorbents in the body with little or no side-effects. Ayurveda also has a socio-cultural context in which family involvement takes precedence over quarantine of the patient. Most of all, its range of indigenous drugs are all tropical and of bewildering variety. They are more suitable for tropical diseases, easily and heaply available for rural masses and present an inexhaustible resource for human cure. Pitted against this profile of Ayurveda, all allopathic drugs as active chemicals are toxic to various degrees with side-effects more severe than the disease itself.

Most of them being synthetic products can be rejected by pathogens by being resistant to them. The diagnostic assessment is quantitative and not qualitative. Western pharmacopoeia become cosmopolitan by appropriating much of Ayurveda by superior technology and standardisation. The medicinal properties of *Chalmugra* oil and *Bel* or *wooden apple* as cure for leprosy and dysentry have already become common factors in both systems. Such possibilities are immense, given the logistic support of modern science and can tackle many diseases declared incurable in allopathy. This is the fourth and final option for free India. Brass while deploring Ayurvedic revivalist chauvinism and preaching integration of Ayurveda within a hegemonic cosmopolitan medicinal system forgets the national self-interest of some of the western countries in swapping Indian indigenous drugs by modern scientific refinement and then patenting tham for national benefit and profit. The recent Dunkel proposal on intellectual property rights is an eye-opener for all integrationists. Leslie on a sober note concludes. If a wizard of modernization decided that 'traditional medicine was an impediment to progress and abolished tomorrow, the whole infrastructure of professionalised Ayurvedic and Yunani practices, he would create a medical catastrophe.[28]

President Radhakrishnan in his foreword to Kutumbiah's Ancient Indian Medicine says: 'The system of medicine will have to keep pace with the developments of time. Our systems suffered because they were not able to reckon with the progress made.' Ayurveda has to live upto this challenge and must be updated and upgraded by modern science and technology without having to sacrifice its world of experience. There is a need to scientifically explore the correlation of ancient insights and modern discoveries beyond the narrow trajectory of purity pollution[29].

References

1. Kutumbiah, P.: Ancient Indian Medicine, Madras 1962: Introduction, C.S, I 30:24.
2. Chattopadhyay, D.P.: Science and Society in Ancient India, Calcutta 1977: 112, 213, 273.
3. *Op.cit.*, Kutumbiah 1962 : 32.
4. Leslie, Charles ed.: Asian Medical Systems : A Comparative Study, London, 1976 "The Ambiguities of Medical Revivalism in Modern India in Leslie ed. 1976; 356-57.
5. Basham, A.L. : 'Practice of Medicine in Ancient India' in Leslie, C. ed. 1976 below : 40.

6a. Bala, P : Imperialism and Medicine in Bengal, Delhi, 1991 : Appendix A. 14-15.

6b. Also-Keswani, N.H. ed.: The Science of Medicine and Physiological Concepts in Ancient and Medieval India, New Delhi, 1974.

6c. Rashid, A: Society and Culture in Medieval India, Calcutta, 1969.

7. *Op.cit.*, Leslie 1976 : 356-57.
8. *Op.cit.*, Bala 1991:41.

9a. *Ibid.*, Bala 1991 : 47.

9b. Ukil, A.ed.: Century Volume of the Medical College, Bengal 1835-1934, Calcutta 1935.

10a. Palit, C: New Viewpoints on Nineteenth Century Bengal, Calcutta 1980: 68-69.

10b. Chattopadhyay, P.: Ayurveder itihas (History of Ayurveda) in Bengali, Cal-1963 :162-63.

11. *Op.cit.*, Govt. Charitable Dispensaries 1840-42 in Bala 1991 : 48-50.

12. Arnold, D. : Imperial Medicine and Indigenous Societies, Mancheater, 1988 'Medical Priorities and Practice in Nineteenth Century British India' South Asian Research Vol. 5 No. 2, 1985 : 160-83.

13. *Op.cit.*, Bala 1991: 51.

14a. Campbell Brown, J.: Report on the Charitable Dispensaries under the Govt. of Bengal for the year 1870, Calcutta 1871.

14b. Carstairs, R.: The Little World of an Indian District Officer, London, 1912 : 154-55.

15. Ray, B.B.: Unish Sataker Banglay Vighan Sadhana (Pursuit of Science in 19th Century Bengal) in Bengali, Cal, 1987 : 97-98.

16. *Op.cit.*, Bala, 1991 : 52-57.

17a. Gupta, Brahmananda : 'Indigenous Medicine in Nineteenth and Twentieth Century Bengal' in Leslie, C. ed. 1976 : 373-74.

17b. *Op.cit.*, Chattopadhyay, p. 1963 : 163-75.

18. Gupta *Ibid.* : 369.

19a. *Ibid.*, Gupta in Leslie ed. 1976 : 371-74.

19b. *Op.cit.*, Chattopadhyay, p. 1963 : 164-75.

20a. *Ibid.*, Gupta in Leslie ed. 1976 : 375-878.

20b. Ghosh Nareshchandra : Ayurved Chikitsa Sastrer Itihas (History of Ayurveda) in Bengali, Cal., 1963 :77.

21. *Ibid.*, Gupta in Leslie ed. 1976; 375-76.

22. Brass, P.R.: 'The Politics of Ayurvedic Education : A case study of revivalism and modernization in India' in Rudolph and Rudolph eds. Education and politics in India, Chicago, 1972 : 344-352.

23. *Op.cit.*, Gupta in Leslie ed.: 374-375.

24. Bhattacharyya, A: Swadeshi Enterprise in Bengal 1900-1920, Calcutta, 1986 : 58-67.

25. *Op.cit.*, Brass in Rudolph and Rudolph eds. 1972 : 351.

26. *Ibid.*, : 370-73.

27. Leslie in Leslie ed. 1977 : 357.

28. *Ibid.*, Leslie in Leslie 1977 : 267.

29. Lele, R.D.: Ayurveda and Modern Medicine, Bombay 1986 : Passim.

10

Epidemics and Empire : A Critique of Public Health Policy in Colonial India

Preamble

A study of the subject has become imperative in view of the growing neocolonial literature depicting India as a dark subcontinent as much as Africa has been branded as the white man's grave. Both are imagined as quagmires of lethal epidemics and decimating diseases. Some of the stereotypes used by British officials in a period of medical imperialism from around 1835 are being recycled from the past by western historians like David Arnold, Ira Klein, Chales Leslie, Paul Brass, Ivan Catanach etc. Roy McLeod, Micheal Worboys, Paul Greenough and J.C. Hume are some exceptions.

Tropical/Temperate

In African Studies today, the dichotomy between tropical and temperate diseases has come to be increasingly challenged. It has been argued that the colonial hangover to regard a group of lethal diseases as 'tropical' is distinctly racist. These diseases like plague, malaria, small pox, cholera, influenza and syphillis are still being held as native to the dark continent or subcontinent. They are assigned to either climatic or genetic causes. Even some typically temperate diseases like tuberculosis and AIDS are being dumped on these countries. It has been contested that some of the so-called tropical diseases have disappeared from the west in the 20th century entirely due to heavy investment in sanitary reform and nutrition. The binary contrast in diseases, in the final analysis, boils down to that between poverty and wealth. The latter can be traced to two hundred years of colonial rule in the third world which caused so much drain of wealth from the colonies to the metropolis (Feierman, Mcleod). It can be maintained that most of these diseases originated in Europe. Plague had been the most dreaded epidemic since the days of ancient Rome. St. Benedict died of it in mid-6th century. Britain was pulverised by plague in the 17th century. It did not abate until the 19th century in some form or other. It has been in Shakespeare as 'plague on both the houses' to describe Elizabethan parliament. It has found its way into the nursery rhyme, 'Ringa Ringa Roses' and as a regular verb of menace in English vocabulary. Malaria (mal=foul, aria=air in Italian) was also of European origin and was rampant in Latin countries from 17th to 19th centuries and beyond. Spain took it to her colonies in Latin America. England suffered from small pox in the late 19th century and the disease was transmitted to settler colonies of

Australia and New Zealand as well as the subject colonies of India. Cholera was chronic in France, in England and the U.S.A. from the mid-19th century. It was carried to Philippines. Russia also had an epidemic of cholera. Syphillis, a dog-borne European disease was carried by seaman into Afro-Asia (Ross, Scott, Snow, Rosenberg, Sullivan, McLeod, Hirst, McNeil). Of the two stories leading to the discovery of the medicinal bark of a wild tree in Peru, one is associated with the illness of the Duchess of Cinchon, wife of the Spanish Governor of Peru who got cured of her fever by taking a concoction of the herb which came to be known as Cinchona after her name. The other story is that the Spanish Jesuits brought it to Europe as a prophylactic to such remittant and intermittent fever (ague, yellow fever, dengue and malaria) and an English apothecary made a trade of it in the 17th century England. Oliver Cromwell would rather die of malaria than take the 'Popish' remedy. Scott lends credence to the first story. It is confirmed by E. Laurie. Italian scientists Grassi and Bignami in 1897 discovered the anopheles female mosquito as the vector of malaria and Italy had no tropical colonies as yet. Alphonse Laveran discovered the plasmodium that invades the bloodstream in 1880. Quinine was in use in French Algiers since 1835.

By then the germ theory of Louis Pasteur had overtaken 'miasma' as the root of these diseases. Yet the tropics were blamed for these epidemics. The heat of the tropical sun and humidity of swampy lands were still held responsible for the diseases. The mosquitoes, lice, rats were condemned as vectors but the root of the virus was not traced. But similar climate prevailed in Latin countries of Europe and parts of U.S.A and England. There was no swampy land nor even a blade of grass in Kalat. Waziristan, the frontier of modern Pakistan yet malaria and kala azar or black water fever was rampant there in the early 20th century. As to lice, rats and cockroaches, the P & O lines carried loads of them along with passengers from Europe to Asia and Australia. Yet when plague came to Bombay in 1896, only Hong Kong connection was blamed. But there was plague epidemic in Karachi and Punjab, Calcutta and Rangoon in the 1910-20 period. It was a shippy disease and one has to turn to Europe for the source. Britannia ruled the epidemics as well. The hosts of the diseases were ignored, only vectors and the tropics were condemned. It is true that the tropical climate was too hostile to whitemen and flared up the germs in their bodies and overpowered their natural resistance in the inter-communicating zone while the local population remained for a while immune in the initial period. But the lethal contact caused a holocaust among the original people of Africa, Asia, Australia, Latin and North America who were left out of the pale of the so-called conquest of epidemics. It was of the whites, by the whites and for the whites. The rhetoric of conquest of disease as a concomitant of territorial expansion. In fact, the first conquered and consolidated more terriotories than gunboats, breechloaders and canonnades. It was medical imperialism par excellence. The Indian Medical Service was a wing of the British Indian Army and catered only to cantonments and actual fields of action. Gradually, it extended to urban areas-civil lines of white population and vested white interests in mines, plantations and factories. The rural areas having the bulk of the native population were left to their fate and demonology. The sanitary measures of public health as advocated by Ronald Ross as the head of the Liverpool School of Tropical Medicine for eradication of malaria in Bengal alone was ridiculed by the official circle as chimerical for the enormous expenditure involved. Ross meekly wanted to apply it only to white installations and living quarters. Even that was turned down though it would be a fragment of the drain of wealth from India. As Mckeown shows, those very measures were the long-term determinants of public health in the developed countries, including U.S.A. It also helped eradicate malaria and yellow fever in Panama, Cuba, Suez, Malaysia and West Africa.

Even if one admits that upto the middle of the 19th century, many such epidemics were global in nature, both tropical and temperate, it is this colossal investment in public health in both sanitary reform and research for prophylactics which made the world of difference between the so-called tropical and temperate zones,-an investment generated by colonial extraction but denied to the colonies. They got the disease but not the remedy. It was indeed a case of colonising body. Cholera is a case in point. Cholera was rampant in France, England, Russia and U.S.A. in the second half of the 19th century, and not just in India on the banks of Ganga. Whereas sanitary reform, especially, water treatment for this water-borne disease eradicated it in the west, nothing similar was carried out in India. Only Punjab derived some such measure after considerable foot-dragging by the bureaucracy. (Scott, Laurie, Headrick, Ross, Megroz, Arthur Dash Memoirs, Curtin, McKeown, Worboys, Snow, Rosenberg, Sullivan, Hume).

Military/Civil

It has already been said in the preceding section that western medicine entered India as military medicine. It was to cater to the cantonments and forward lines to win battles of expansion. The main enemies were not the native princes alone but the diseases and epidemics which decimated the British Army set up in the early 19th century fought this part of the more lethal battle and it can be said that medical imperialism actually consolidated the empire. The IMS fought kala-azar in Waziristan in the port cities of Bombay, Karachi, Calcutta and Rangoon and kept the flow of trade and commerce unabated. In the plains, it grappled with malaria, cholera, small pox. None of them were tropical diseases per se. Most of them were European in origin and some global in character given the state of public health management in the mid-19th century. They had nothing to do with tropical miasma. Yet western histories are replete with stories of victory over tropical diseases and triumph of empiricism. They project the tropics as the contemptible other, chockful of malignant diseases. The so-called victory was not the victory over the disease as a whole as should have been the criterion of medical science. The whites contaminated the native population with their lethal contact and left them economically impoverished in the grip of malnutrition and epidemics without extending equal treatment to the occupied. The victory was in a narrow sense over the inherited diseases of the whites in a hostile climate. The dark continent and the subcontinent remained the white man's grave for a long time till they became the graveyard of the native population. This was achieved through a dual policy of drain of wealth and health of the subject-nations. The dialogue between epidemics and empire has been hopelessly one-sided. It has been only the success story of the hegemonist. Though Headrick calls western medicine one of the tools of empire, he does not bring out the full implication of medical imperialism. He is also taken up with the success of metropolitan medicine and has no space for its fall out on the periphery. Yet Mcleod in his superb introduction to Disease, Medicine and Empire queries this smugness. In a stunning book, Kenneth Ballhatchet exposes this hypocrisy. In dealing with a delicate subject like the immoral traffic between the army and the prostitutes, he narrates how lustful white soldiers had to be provided with prostitutes in the red bazaars around the cantonments and the VDs were blamed on the loose women alone. Contagious Diseases and Cantonments Acts were passed in the 1860s to register prostitutes, carry out their health inspection and put them in lock hospitals forcibly for treatment to safeguard army. The vector of VD had to be the hapless professional without a thought that the white soldier could also be its carrier as he was of syphillis, a European disease. Yet within the purview of army requirement, the lock hospitals provided the minimal treatment to the abused women. Even this was gradually withdrawn after 1872 raising the incidence of the VD phenomenally again. Lal bazaars apart, the orderly bazaars skirting the barracks as in Barrackpore were vulnerable to European contact and contagion without any scope of treatment.

The civil native population were left out of the scope of public health service. A section of the IMS personnel was periodically transferred to man the civil hospitals in the presidency towns and district headquarters. They mostly attended white population in urban areas and the precincts of plantations, mines and factories. The military duty of the IMS was of such priority that even within the white cadre, there was strong grievance against it. Ronald Ross in the middle of intense research on mosquitoes and malaria at Hyderabad, was subjected to hectic transfer to N.W. Frontier, Rajputana, Burma, Madras and Calcutta affecting his work. It also affected other members like Laurie who was engaged in his research in kala-azar in Waziristan but transferred to Calcutta. The masses of the civil population who lived in the rural areas remained virtually beyond the purview of western medicine.

Urban/Rural

The urban set-up of the IMS for the civil population was itself peripheral to the military establishments. A few metropolitan hospitals in the district headquarters were meant for the urban elites and there too, the white ruling class was privileged. Besides, the civil surgeon (white) and the Indian sub-assistant surgeons in the district headquarters who did a lot of private practice were mostly unavailable for rural masses. The government virtually disowned them. There was no public health department nor any budgetary provision for sanitary reforms even in periods of malaria and cholera epidemics. Only in 1920s when the Local Self-Government Act was passed and District Boards and Union Boards took over the responsibility and bulk funding of rural health, the district dispensaries in rural areas were set up. But they were ramshackle huts with few medicines and barefoot doctors trained in the medical schools. More often the local vaids and hakims were pressed into service.

The indigenous systems of medicine were discouraged and discontinued which were both cheap and readily available at the doorstep without allopathic replacements. The vaids and hakims were castigated as quacks depending on folklore and no exact science though they were the ultimate resort of the villagers. The epidemics of malaria, cholera, small pox and plague were brought to the countryside by government policy of immigration, railway transportation and embankments. White-owned plantations, mines and factories herded colonies of coolies (labour) in the cooly lines of dingy barracks spreading the contagion. Many of the diseases were spread from the ship to the field in this way (Ramasubban). In the district gazetteers and settlement reports, it is clearly pointed out that some of the districts like Jessore, Hooghly and Bankura in Bengal which were known as healthy were becoming death traps over the 19th century. The rural masses did resort to their gods and goddesses like Sitala and Olabibi for moral and mental reinforcement against the mighty epidemics but they were not culturally averse to allopathic medicine and treatment. From the district magistrate's tour diaries, it is evident that villagers travelled many miles and stood patiently in queue for a phial quinine or inoculation. But these were in chronic short supply. The government clearly had no infrastructure for rural health care. While epidemics were transported to the interior, there was no remedy. Traditional medicine had been languishing without patronage and its practitioners had not been incorporated in the health care programme. The villagers were allegedly religious obscurantists and this cultural opposition stood in the way of the introduction of western medicine. Rumour-mongering against official doctors and medicine by villagers made it difficult for the government to penetrate villages. This official escapism has been harped only Ivan Catanach, Arnold Klein. The truth of the matter has been spelt out by Frantz Fanon and Steven Feierman in the African context. It was the question of trust between the occupier and the occupied which had been destroyed by centuries of exploitation and foul play. Going to see a doctor was thought as hazardous as going to the police.

(Tour diaries of Bell, Donovan, Bengal civilians, R. Nicholas, D. Arnold, Ira Klein, I. Catsnach, Frantz Fanon, Steven Feierman)

Segregation/Community Healing

Segregation was the watchword of imperial medicine. It was primarily a racist outlook which treated the indegenous population with contempt for having filthy and unhygienic habits and being vectors of all kinds of infectious diseases. Cantonments and civil lines inhabited by the whites were generally separated from the native dwellings. The duality of white towns having civic amenities and black towns having none of it was typical of British India. Yet most of the epidemics spread from the ports and cities inhabited by white sailors, soldiers, merchants vagabonds. It has been shown in a previous section that the white town still needed their lal bazars and orderly bazars for service, sexual and manual. In the event of diseases following such intercourse, the indigenous people were blamed and segregated. Lock hospitals were not only for the prostitutes but all city hospitals became virtually lock hospitals during epidemics like the plague which gripped Bombay from 1896 and Calcutta in 1912.

As plague threatened not only civilian and military life but also international trade in cotton, oilseeds and grain due to panic among the European governments, the British officials in India were frenzied over the epidemic. There was blatant and brutal intrusion into the private houses of the native population on mere suspicion of disease. Their hearths and homes were virtually demolished, floors dug up and D.D.T sprayed, walls cracked up and limewashed and any inmate suspected of the disease was pulled up and carried in vans to the infectious diseases hospitals for quarantine. They were cut off from their relatives and many died on their way to the hospital. Their edad bodies were not even handed over to the relatives for cremation. It was nothing but ruthless segregation without treatment. It was clear that the authorities only wanted to safeguard themselves and did not care to save the lives of the unfortunate victims. There was no compassion, care or empathy essential for healing.

This was a culture shock of a different kind born out of enormous credibility gap and not religious and cultural obscurantism. Traditional medicine and medical care depended on community healing and family care. The patient was treated in body and mind in his personalised circumstances. The spread of rumour against government intentions was not due to any fundamental cultural opposition. It was due to lack of credibility and treatment backed by brutal intervention and segregation. The plague commissioner Rand was murdered in Bombay because of his excesses. Benign intervention and segregation in cases of infectious disease would not be challenged. It was the dehumanised treatment of the patient which was resented by the local people. Hence the rumour of water-poisoning and outright poisoning of the patient during the plague epidemic in Bombay. Panic and rumour were not necessarily rustic propensities. The authorities also suffered from the same and took drastic action. In Bengal the rural people came from a long distance for cholera vaccination when mutual trust was built up between the government and the governed.

(Arnold, Klein, Catanach, Chandavarkar, Nicholas,Donovan and Dash memories Frantz Fanon, Feierman).

Apart from racism, western medicine by itself was concerned with the etiology of the disease by the limit or the vital parts repaired. The whole concept is based on segregation of the patient from the community and body from the mind. The total perception of the patient as a human being, a member of a family and community and having an interactive mind is missing in it. But these considerations are important in effecting a cure. It is being increasingly recognised in the west that almost all diseases are psychosomatic, needing community or at least family care. (John Ehren-reich, Dubos, Selye, J. Cell).

Traditional/Western Medicine

Before the medical intervention of the British, both Ayurvedic and Yunani systems of medicine were prevalent in India. Though traditional medicine was not so much institutionalised, it was fairly professionalised and there existed schools of vaids and hakims founded by stalwarts like Gangadhara Roy in Bengal and Hakim Ajmal Khan in Delhi during the mid-19th century and later. The standard Arabic and Sanskrit texts were read and acted upon in these schools and the British authorities encouraged them in government institutions like the Benares Sanskrit College, Calcutta Sanskrit College and Native Medical Institution in Calcutta till 1835. But then there was the end of paternalism and government patronage was diverted to western medicine and medical colleges set up in metropolitan cities. With the consolidation of British rule and achievements of western medicine at home, traditional medicine was dismissed as folklore and western medicine was the only stream taught at the Medical College and practised in government hospitals and dispensaries. Meanwhile traditional medical texts were translated and ransacked and the pharmacopoeia was appropriated in the British pharmacopoeia and many traditional herbal medicines were converted into western drugs by chemical analysis and standardisation. The indigenous vaids and hakims were still recruited as barefoot doctors to man medical care in the countryside after some training in medicine.

Traditional medicine was undoubtedly effective for elemental chronic diseases born out of imbalance of wind, biles and caugh and even in many other viral diseases of the tropics which were much less in appearance than the later epidemics imported from the west. It was not given a fair trial by suitable encouragement and allowed to exist by sufferance in the rural areas where western medicine never reached. The treatment was personalised and cheap and there was a good rapport between the vaid and the patient. Later the Indian practitioners founded their own institutions and formulated an integrated course having the best of both traditions. Drugs were also produced accordingly. Still there was no official recognition and all public health expenditure was lavished on western medicine alone. Against this derecognition there was considerable politicisation of the issue and regional and national ayurvedic congresses were formed to press its case on merit. But the withering touch of colonial rule did not allow it to grow even in the vast rural sector which was denied medical care altogether. The solution lay not in the absorption of traditional into the hegemonic regime of western medicine to be euphemistically called 'cosmopolitan' medicine but the development of an alternative medical system standardised by the logistics of western laboratories as it is being done in free India today. Obviously, microbial diseases, especially those coming from the west could not be tackled by traditional medicine but still it had efficacy in many other diseases for which there was two western remedy. (Leslie, Brass, Bala, Palit).

Eastern Sounding Board for Western Medicine

It has already been observed that the empire was half won in medical laboratories and empirical work in the empire itself. It was a life and death question for the imperial army against the invisible enemies, the microbes in full play under the tropical sun. The tropics have been variously described as the white man's grave or the process of east-west encounter as migration to death. If diseases and epidemics could be surmounted, the army would be in full strength and military battles could be won. It was medicine for empire or medical imperialism at large. Medical research was directed to the conquest of 'tropical' diseases among the army (mainly white) and next the white ruling class. Thus, Ronald Ross was engaged in research in mosquito-borne malaria at Hyderabad and Robert Koch in tuberculosis and cholera at Calcutta. Macnamara and Rogers followed up research in cholera at Calcutta, Haffkine worked on plague at Bombay and Laurie on kala azar in Waziristan. Take the example of Ronald Ross, Ross, an IMS officer began his research at Hyderabad and ended it at Calcutta at the end of the 19th century. He went on mosquito-keeping, feeding them on malarious patients and healthy poor people who volunteered to be the guinea pigs on pittance alternately. When he was transferred to Calcutta he found it difficult to get such volunteers. So he carried on the experiment with cageful of parrots until he located the bacteria in the vector mosquito (female anopheles variety). The theory that was propounded by Manson, Laveran, Grassi and Ross himself was thus experimentally proved and he received the Nobel Prize for this discovery. Its therapy was already known in Europe and its other colonies in Africa, Asia and Latin America. It was the graduated dose of quinine, an extract of cinchona naturally grown in Latin America and successfully transplanted in Indonesia and India by the Dutch and the British authorities. It was finally synthetically produced from the same alkaloids during the Second World War. Ross's discovery led to the other measure : sanitary reform on war-footing by clearing of jungles and swampy areas and killing of mosquitoes by sprays of D.D.T. and petrol in ponds, canals and embankments forming pools of rainwater. The Liverpool School of Tropical Medicine that he formed and directed, carried out commissioned medical expeditions to Malaysia, West and South Africa, Suez, Panama and Cuba where scheduled sanitary reform was undertaken at a huge expense more than compensated for by gain in life in the pantations. The entire west benefited from Ross's research, most of all, Britain but when he suggested that such reforms be carried out in at least Bengal, the authorities ignored it on grounds of expense. Even his suggestion for urban areas, plantations and mines was turned down despite so much drain of wealth and health from India as a whole and Bengal in particular. Only the quinine industry thrived, first the natural produce and then the synthetic product while malaria raged in village India. (Megroz, McKeown, Worboys).

Public Health/Private Practice

We have already seen that colonial experience was put to good use in other parts of the empire of Britain and other western countries. Britain was cured of most of the fatal diseases due to heavy government investment in public health programmes of sanitation and nutrition. But when it came to implementing it in India, the government dragged its feet on grounds of expense. Sanitary reform was relegated to a secondary position and germ theory and prophylactics made overriding considerations in official policies aided and abetted by the London School of Tropical Medicine headed by Manson. Hume shows convincingly in the case of Punjab how a successful plan of public health with the help of a brigade of vaids and hakims achieved wonders and the effort to prevent water pollution in the wake of similar reform in London after its connection with cholera epidemic was convincingly proved by John Snow saved the region from cholera epidemics. But this was gradually obfuscated and abandoned in favour of the doctor and his clinic. In Bombay, it was the same before the plague epidemic. There was a dualistic development of the city having concrete jungles skirted by slums. The civic amenities were confined to the business district and the administrative enclaves. Public health was not the concern of the government. It was left to the private practice of the government. It was left to the private practice of the fee-fleecing doctors and cartels of chemists and druggists. If this was not satellization, what else was? Yet Ira Klein argues to the contrary. He blames the rising commercial middle class for the haphazard growth of the city and makes public health a private responsibility, of the citizens. The government only leapt into brutal intervention during the plague epidemic when it affected international grade in cotton and grain. Private practice only fattened the purse of the European doctors and Indian practitioners of western medicine in urban areas. Privatisation of public health in a poor country of rural masses was denial of equal treatment. McKeown has demonstrated from vital statistics in Britain that nutrition and hygiene more than therapy and prophylactics had been responsible for the decline of mortality there. Yet this essential aspect of public health was never given a fair trial in India. Instead, traditional medicine was dismissed as folklore and indigeneous vaids and hakims ridiculed as quacks, thus sapping the only possible alternative medical care by sheer neglect. A proper Public Health Ministry came to function only after 1920 with the passing of Local Self-Government Act when people in the interior had to pay various cesses for their health care and after 1937, the portfolio passed into Indian hands. (Hume, Snow, Klein, Palit).

Western Drugs : Eastern Medicine

The cheap, locally procurable indigenous medicine was already languishing due to lack of patronage. It was being increasingly replaced by western drugs from the 1880s when they could be mass-produced and tableted at a cheap rate. The vast area of medically unprotected hinterland was left to the imperial drug industry to flood with pills, capsules and syrups. It was a bonanza for the private practitioners and chemists of western medicine. It was medical imperialism of another kind. In the monopoly market, many experimental drugs were pushed to test their prophylactic power at the cost of humanity. Such a human laboratory was nowhere to be obtained. The refined drugs were retained by the west and the rejected ones known for their toxicity and other side-effects were dumped by multinational companies till independence. It continues even now through neocolonial devices of patenting and monopoly.

(Bala, Samaddar)

References

1. David Arnold, Colonising Body : State Medicine and Epidemic, Barkeley, 1993 ed., Imperial Medicine and Indigenous Societies, Manchester, 1988.

2. Ira Klein, 'Death in India' in Journal of Asian Studies, 32.4 (1973); 'Malaria and Mortality in Bengal' in Indian Economic and Social History Review 9.2 (1972); 'Urban Development and Death : Bombay City, 1870-1914' in Modern Asian Studies 20.4 (1986); 'Plague, Policy and Popular Unrest in British India' in Modern Asian Studies 22.4 (1988).

3. Charles Leslie ed., Asian Medical Systems : A Comparative Study, Berkeley & Los Angeles, 1976 : Articles by A.R. Beals, B. Gupta and himself.

4. Paul Brass on Ayurvedic Education and Politics in L. &. S. Rodolph eds. Education and Politics in India, Chicago, 1974.

5. Catanach, I. 'Plague and the Indian Village, 1896-1914' in P. Robb ed. Rural India : Land, Power and Society under British Rule, London, 1983, 'Plague and Tensions of Empire, 1896-1918' in D. Arnold ed. Imperial Medicine cited; 'Poona Politicians and the Plague' in J. Massalos ed., Struggling and Ruling : The Indian National Congress 1885-1985, New Delhi, 1987.

6. Roy McLeod & Milton Lewis eds., Disease, Medicine and Empire : Perspectives on Western medicines and the Experience of European Expansion, London and New York, 1988.

7. M. Worboys, 'Manson, Ross and Colonial Medical Policy: Tropical Medicine in London and Liverpool, 1899-1914' in *Ibid.*

8. P.R. Greenough, 'Varolation and Vaccination in South Asia, c 1700-1865' in Social Science and Medicine 14: 345-47.

9. J. C. Hume, 'Colonialism and Sanitary Medicine : The Development of Preventive Health Policy in the Punjab 1860-1900' in Modern Asian Studies 20.4 (1984).

10. S. Feierman, 'The Struggle for Control : The Social Roots of Health and Healing in Modern Africa' in African Studies Review 28 (2/3), 1974.

11. Ronald Ross, Autobiography, London, 1910.

12. R. L. Megroz, Ronald Ross : Discoverer and Creator, London, 1931

13. H.H. Scott, A History of Tropical Medicine, 2 Vols, London, 1939.

14. John Snow, Snow on Cholera, Reprint, New York, 1936.

15. C.E. Rosenberg, The Cholera Years : The United States in 1832, 1849 and 1866, Chicago, 1962.

16. R.E. McCrew, Russia and the Cholera, 1823-32.

17. R. Sullivan, Cholera and Colonialism in the Philippines, 1899-1903.

18. L. F. Hirst, The Conquest of Plague : A Study of the Evolution of Epidemiology, Oxford, 1953.

19. R. Pollitzer, Plague, Geneva, 1954.

20. W. H. McNeill, Plagues and Peoples, Garden City, N.Y., 1976.

21. D. H. Headrick, The Tools of Empire : Technology and European Imperialism in the nineteenth century. N.Y. 1981, ch. 8.

22. W. Laurie, Blackwater Fever in Waziristan, 2 vols., 1938-Mss. in the Centre of South Asian Studies, Cambridge, U.K.

23. A.J. Dash, ICS, A Bengal Diary, Vols. 1-9, 1910-47, cf : chapters 2, 4, 6.

24. Philip Curtin, The Image of Africa : British Ideas and Actions 1780-1850. 2 Vols., Madison, 1964, Part III, Death by Migration : Europe's Encounter with the Tropical World in the 19th century, Cambridge, 1989.

25. Thomas McKeown, The Modern Rise of Population, London, 1976.

26. K.A. Ballhatchet, race, Sex and Class under the Raj N.Y., 1980.

27. Radhika Ramasubban, 'Imperial Health in British India' in McLeod ed. Imperial Medicine.

28. J.M.G. Bell, ICS, Tour Diary, 1941-47 mss in the Centre of South Asian Studies, Cambridge.

29. J.T. Donovan, ICS, Tour Diaries 1927-31 mss in *Ibid*.

30. R. Nicholas, 'The Goddess Sitala and Epidemic Small Pox in Bengal' in Journal of Asian Studies 61.1 (Nov., 1981).
31. Frantz Fanon, 'Medicine and Colonialism' in J. Ehrenreich ed. Cultural Crisis in Modern Medicine, N.Y. 1978.
32. R. Chandavarkar, 'Plague Panic and Epidemic Politics in India 1896-1914' in T. Ranger & P. Slack eds. Epidemics and Ideas : Essays on the Historical Perception of Pestilence, Cambridge, 1992.
33. Rene Dubos, Man, Medicine and Environment, N.Y., 1968.
34. Hans Selye, The Stress of Life; N.Y. 1976. Revised Edition.
35. J.W. Cell, 'Anglo-Indian Medical Theory and the Origins of Segregation in west Africa' America Historical Review 91.2 (1986).
36. Poonam Bala, Imperialism and Medicine in Bengal, New Delhi, 1991.
37. C. Palit, 'From the Raj to the Panchayati Raj : Local Self-Government in Bengal 1920-50' in Palit ed. Political Economy and Protest in Colonial India (forthcoming). 'Ayurveda in Colonial Bengal' in the Souvenir of the International Seminar on Traditional Medicine sponsored by the Indian Pharmacological Society, Calcutta, 1993.
38. R. Samaddar, 'Health Administration in India since 1947-a critique'-socialise perspective, Vol. 13, November, 1987.

11

Ronald Ross and Malaria

Introduction

Calcutta is celebrating the centenary of Sir Ronald Ross's discovery of the mosquito cycle of malaria parasite (Plasmodium falciparam).

While working in the P.G. Hospital here in 1897. This discovery earned him the Nobel Prize in 1902 and in this sense Ross was Calcutta's first Nobel Laureate. In fact Ross had more than one Indian connection. His father Campbell Ross was in the Indian Army and rose to be one of its Brigadier generals before retirement. Ross was born at Almora in Kumaon (U.P.) in May 1857, joined the Indian Medical Service and served India till 1897 and is known as a friend of India.[1] Ross is therefore rightly remembered by India on the occasion of the centenary of his discovery. Now that malaria has returned with virulence,[2] the occasion gives us an opportunity to ponder over his discovery and prescription and what went wrong since his departure.

Ross in I.M.S.

Ronald was devoted more to literature and was already writing poems when his father decided to put him to St. Bartholomew's medical school. His elder brother was already in the Indian Army but his father decided that he should join the I.M.S. which was then a safe and lucrative job. His father had suffered from bouts of malaria during his tenure in India. He had a relapse of tertian malaria in England several years after his retirement from service. This might have also been a reason for him to persuade his son to join medical service and fight malaria. Ronald passed his medical examination without difficulty and duly joined the Indian Medical Service gifted as he was in many fields of knowledge.[3] By then he had already published one or two books of poems .... Megroz, his biographer considers him to be a major poet as a proof of which one of his poems is cited below:-

Indian Fevers

In this, O Nature, yield I pray to me.

I pace and pace, and think and think, and take

The fever'd hands, and note down all I see.

That some dim distant light may haply break

....

The painful faces ask, can we not cure?

We answer, no not yet; we seek the laws

O God, reveal thro' all this things obscure

The unseen, small, but million murdering cause[4]

His first posting in India was in Madras in 1884. He records in his Memoirs that he had the first experience of vigorous mosquito bite in the mess where he was putting up. After attending his hospital duties, he continued to write poetry in his spare time and send them to literary magazines in England for publication. He was also interested in mathematics and according to Megroz, some of his theorems were enough to establish him as a genius in the field. He also invented and constructed his own microscope for his study of the microbes.[5] He was soon transferred to Bangalore and there he had another brush with mosquito bite. Looking for the mosquitoes from his bedside window, he found the flower pots on the parapet which he thought to be the breeding ground of the mosquitoes. He got this pot removed and had instant relief from the attack of mosquitoes. This set him thinking on the connection between mosquitoes and malaria[6] he had to treat in the hospitals. He gradually gave up his passion for poetry and mathematics and concentrated more on mosquitoes as a vector of malaria. This was the time when considerable research was done on parasites in the blood by Alphonse Laveran of France, who did his research at Bone in Algieers from 1876 onwards. In Italy a whole team of parasitologists led by Bignani, Marchiafava, Goli and Celli were all busy in locating a malaria parasite in human blood. Malaria or Mal (foul) aria (air) was widely known in Italy to be a disease caused by miasma from rotting vegetation. This was being disproved by the discovery of the parasite.[7] Patrick Manson, a British Doctor posted in Amoy in China about this time had already proved that filaria was caused in a healthy person by mosquitoes as a vector after biting an infected body. He had also proved that malaria was caused by the presence of a parasite in the bloodstream of unaffected persons in his correspondence with Ross. He had admitted that mosquito could be the carrier of this parasite. In this background Ross began his intensive research on the relation between mosquito and malaria at Secunderbad from 1891 onwards.[8]

Research at Secunderbad

In between his duties at the military hospital, Ross got engaged in his 'Great Problem' i.e. connection between mosquito and malaria. He became a regular collector of mosquitoes, particularly Anopheles female, the species he pinpointed as the carrier of the parasite. He kept jars full of mosquitoes and whenever he could, dissected them to find out the location of the parasite in the mosquito's stomach and put them under his microscope for clear observation. Sometimes it yielded valuable results and sometimes the effort was in vain. Abdul, his valet was forever engaged in catching the mosquitoes and putting them in jars for his master. He carried on his experiments with them. Whenever Ross found the parasite in the stomach of the mosquito, he watched the physical nature of its development and recorded it. Some time he would release the carrier mosquitoes on the malarious patient and then catch those mosquitoes back after their biting for further observation.[9] He would hire men from the bazar and pay them a rupee each for a prick to test the blood of a possible infected person. His other bearer Luchhman was made a willing partner in one of his experiments in which he was made to drink a glass of water full of dead and carrier mosquitoes to find out whether he had the fever. Luchhman had the fever as a result but quinine was always kept ready as a safety valve.[10] The higher authorities never encouraged him and infact dissuaded him by giving him more hospital duties and ordering transfer to Burma, Rajputana and Baluchistan intermittently during the throbbing years of the nineties.[11]

On a furlough to England in 1895, he met Dr. Manson and daily discussed his findings with him. Manson was not convinced of what he was talking about, i.e., metaxeny or the growth and development of the malaria parasites in the stomach of the mosquitoes after it had sucked the blood containing the parasite from a patient. Thus the mosquito was playing the second host of the parasite. It was then transmitted to a healthy person's bloodstream by the carrier mosquito causing the diseases by affecting blood circulation and growing in the serum of the second host.[12]

In a series of fifty letters written half and half by Ross to Manson and vis-a-vis, Ross continued to report his research and seek advice from Manson, who still had doubt about his discovery.[13] These letters constitute the memoir itself. Some specimens are quoted below :-

(1) Dear Dr. Manson Secunderabad, 15 May 1895.

The fortnight since I wrote to you has been made up partly by disasters and at last by success. I had a bad case requiring constant attendance; the two ague cases in my hospital ceased to show parasites, when, with much difficulty, I had got two cases out of the bazaar, all my mosquitoes died again; and next day when I had got a fresh crop of mosquitoes my two cases ran away, because I pricked their finger (sic.) in spite of my giving them a rupee a prick.

This last experiment then seems to me to prove that the crescents undergo change into spherules in the stomach of the mosquito as outside; and to suggest that they do so faster than in control specimens. I must satisfy myself much more however before I state so much definitely. I am as it were touching the matter with my soul before my intellect has reached it; just as a phagocyte touches a microbe with its bioplasm while the nucleus lumbers up behind.

Yours truly,

R. Ross[14]

(2) My Dear Ross

21 Queen Anne Street
Cavendish Square
17th May, 1895

I was delighted to get your letters from Malta and from Secunderabad and to gather from them that you pursue the plasmodium with enthusiam. You do not lack for material and energy, and though you may get many a check. I firmly believe you will work out some important result; you can not fail to do so.......... Already you have proved that crombie is wrong and that a diagnosis of malaria can be made with the microscope in less than five minutes. You must set them right about this in the first place.

Soon I expect to hear that you have followed the plamodium into the mosquito or some other suctorial insect. If you succeed in this you will go up like a shot and get any facilities you may ask for in the way of leisure for investigation and laboratory appliances and assistance. Hoping to hear again from you soon, and assuring you of my sympathies in your work, and willingness to help you in any way I can.

I remain,

Yours very truly,
Patrick Manson.

In these letters, Ross tries to convince Manson about the mosquito cycle of malaria but Manson still stuck to his view that mosquito was only the inoculator of the parasite and not its host. Ross decided on further investigation but he was soon transferred to Calcutta in 1897. The higher authority was throughout opposed to his research beyond his official duties in the hospital; hence the transfer. But Ross was indomitable. While posted in the Presidency General Hospital in Calcutta, he continued his laboratory work on mosquitoes. This time his assistant was Muhammad Bux who would catch the mosquitoes for dissection, Ross experimented with pigeons, crows, larks, sparrows and other birds by releasing carrier mosquitoes on them. Even birds contracted malaria by this process.[16] After adequate experiments Ross wrote down the results of his research. He was able to prove beyond doubt that mosquitoes were carriers of malaria parasite and also a vector, not only on human bodies but also birds. He sent this historic paper to England which earned him a Nobel Prize in 1902.[17] While doing this valuable paper work, he was sent to Assam to investigate Kalazar epidemics there. His research on malaria at Calcutta had to be kept in abeyance for some time. He wrote an interesting report on Kalazar which to this day remains a pioneering work in that field. Despite such disturbances, he resumed his experiment on mosquitoes and malaria in Calcutta and wrote back to Manson about the results thus,

(3) Dear Dr. Manson 8, Harrington Street, Calcutta.
10 March 1898.

On the pigmented cells again to-day (in one mosquito) and have found active vermicules in mosquitoes recently fed in a crow......

To-day (15th) I continued examination of mosquitoes fed on all the birds three days ago (11-12th). All except one were negative. That one contained two pigmented cells. Of course I don't know which of the birds it bit, but fancy it was the crow. The pigmented cells were in the stomach (I glance at all the tissues formally). One was about 15μ and the other 9μ each containing about 10 granules of pigment (drawing).

With kind regards.

Yours truly,
R. Ross.

It is surprising to note in his letter that there was little malaria in Calcutta at that time and to find out infected persons for experiment was very troublesome. That was one of the reasons for his training the experiment on the birds. But it was also to prove that human body was not only the host of malaria parasite. Mosquito was thus singled out as the incubator of Plasmoduim falciparam. Calcutta thus provided him with the last Laboratory and aviary for his epoch-making final discovery of a certain connection between mosquito and malaria.[19] Hence he should be regarded as Calcutta's first Nobel Laureate.

Medical Missionary

Ross was not content with the discovery itself. He was convinced that the Government of India must undertake sanitary reforms for eradication of malaria.

Mosquitoes must not be allowed to grow in unclean water bodies, polluted vegetation and wet lands. He wrote with W.G. King to the Government of India appealing for funds for such a campaign out of India's revenues.

It was estimated that mortality and debility of India's labourers caused by malaria epidemic reduced productivity so much that if malaria was contained that would compensate for the expenditure of sanitation. But the Government of India did not pay heed to his appeal. Even Manson, his friend philosopher and guide, who was then appointed medical advisor to the Secretary of State for India advised him to confine himself to diagnostics and treatment and desist from campaign for sanitary reforms.[20] Ross was recalled to England in 1899 and was made the Director of the School of Tropical Medicine of Liverpool. On the recommendation of Manson who was already the Director of the School of Tropical Medicine in London. This was no doubt a personal elevation and reward for Ronald Ross for his life-consuming research but India lost a friend, who was espousing her case in right earnest. Ross was removed from the scene of his research but his missionary zeal was unabated. As the Head of the Liverpool School, he was able to carry out his ideas of sanitary reform in the colonies of imperial Britain. He and his colleagues led medical missions to Panama, Suez, Congo and Malaysia to carry out sanitary reforms on a large scale in these malaria-ridden territories. Very soon, these places were freed from malaria and became highly profitable colonies for the mother country. The London-Liverpool Controversy over eradication of malaria remained in the background throughout. London under Manson still insisted that malaria only needed clinical treatment and had not much to do with sanitary reforms, but Liverpool under Ross carried out the medical missions with thumping success proving Ross's discovery to the hilt.[21]

Conclusion

In conclusion it must be said to the credit of Ronald Ross that he was a genuine friend of India. It was he who repeatedly wrote in his papers on malaria and also in memoirs that malaria was a tropical disease spreading all over the world from Latin America to South East Asia and also to northern countries of Europe including England in contemporary Europe, when Europe lacked proper sanitation and nutrition. It was therefore not a disease of the dark subcontinent. It was confirmed by H.H. Scott in his 'A History of Tropical Medicine' though still assumed by fashionable medical historians like Ira Klein, Philip, Curtin and David Arnold. Ross is evidently freed from such white prejudice.[22] Ross in his memoirs repeatedly acknowledges the contribution of Indian people and particularly his valet who had ungrudgingly cooperated with him in all phases of his research. India had been throughout the colonial period the sounding board of all kinds of scientific experiments and even theories of social science for perfection and later use in the metropolis and the empire. Ross was conscious of his debt to India as much as India was indebted to him. Malaria has come back with a virulence to Calcutta, a historic city which hosted his epochal research and ultimate findings. It behoves us not only to revive his memory by maintaining the heritage building where he worked but also to apply his sanitation programme on a war-footing to rid the city of mosquitoes which brought him so much success during the medical missions to the tropics launched by Liverpool. Let us end by quoting the poem that is still inscribed on a tablet in the precincts of the P.G. Hospital,

This day relenting God

Hath placed within my hand

A wondrous thing! And God

be praised at his command

seeking his secret deeds,

with tears and toiling breath

I find the cunning seeds

O, million-murdering Death.

I know this little thing

A myriad men will save.

O Death where is thy sting,

The (sic.) Victory, O grave?[23]

References

1. R.L. Megroz : Ronald Ross : Discover and Creator London 1931, p/31; Ronald Ross : Malarial Fever New York, 1902, p-10, Sandeep Sinha, Public Health Policy and the Indian Public Bengal, 1850-1920, Calcutta, 1998, passim.
2. P.G. Bulletin, Issue 3, January, 1988, Calcutta p. 3.
3. R.L. Megroz, *op.cit.*, pp. 31-39.
4. *Ibid.*, p. 50.
5. Ronald Ross-Memoirs, London, 1923, p. 152.
6. *Ibid.*
7. R.L. Megroz, *op.cit.*, pp. 51-54.
8. Ronald Ross *op.cit.*, p. 132.
9. *Ibid.*, p. 156.
10. *Ibid.*
11. *Ibid.*, pp. 179-240.
12. *Ibid.*, p. 223.
13. *Ibid.*, p. 153.
14. *Ibid.*, pp. 139-142.
15. *Ibid.*, pp. 152-153.
16. R.L. Megroz, *op.cit.*, p. 71.
17. Ronald Ross : Memoirs, *op.cit.*, p. 479.
18. Ronald Ross : *op.cit.*, pp. 267-269.
19. P.G. Bulletin *op.cit.*, p. 1.
20. Michael Worboys' article in Roy Mc lead and M. Lewis edited: Disease, Medicine and Empire, London, New

York 1988 passim; Ronald Ross: Report of the Malaria, Expedition of the Liverpool school of tropical medicine, Liverpool 1900, p. 6 C.A. Bently : The English Sanitary Reformation, Calcutta, 1925, p. 72.

21. *Ibid.*

22. H.H. Scott : A History of Tropical Medicine, London 1939 Vol. 1, Introduction C. Palit, 'Epidemics and Empire' in C. Palit (ed.) Revolt Studies Vol. 3, Calcutta, 1998. Passim.

23. P.G. Bulletin *op.cit.*, p. 1.

SECTION-IV : ENVIRONMENT

12

British Economic Penetration into the North Eastern Hills : Overland Trade and Allied Questions, 1800-1850

The Anglo-Nepalese War of 1814-16 is conventionally described as a British move to contain Gurkha aggression on the British India border territory of Basti and Gorakhpur.[1] A deeper probe has, however, revealed British designs on Nepal under Gurkha rule since 1768. The East India Company intended to receive fabled Nepal trade and open up China through Nepal and Tibet. Kinloch and Logen expeditions had tried to subvert Prithvinarayan's Gurkha rule by effectively siding with the Newar Raja Joyprokash Malla.[2]

The British attempted systematic penetration into the eastern sector of the Himalayan trade during the time of Warren Hastings. Hastings in a minute of 19th

April, 1779 had outlined his long-term northern policy : 'Like the navigation of unknown seas which are exploited not for the attainment of any certain or prescribed object but for the discovery of what they may contain; in so remote a search, we can only propose to adventure for possibilities.'[3] Hastings planned the missions of Bogle (1774 & 1779), Turner (1783), Purangir Gossain (1785) to Tibet and Hamilton (1775) to Bhutan. Markham writes : 'The opening of the road through Nepal and obtaining the abolition of the duties and exactions which have lately been imposed on trade in that country appears an object of great importance towards establishing a free communication between Bengal and Tibet.[4]

His policies were kept up by Cornwallis who through the instrumentality of Jonathan Duncan, Resident at Benares, secured a commercial treaty with Nepal on 1st March, 1792 which agreed on a reciprocal duty of $2^1/_2$% on exports and imports between India and Nepal and indemnity in case of robbery and theft. The treaty was silent on trade with Tibet. Nepal had wooed British assistance in view of Nepal's war with Tibet. But the British would not side with Nepal and be rather friendly with Tibet. Nepal broke off ties and however, Captain Kirkpatrick, Cornwallis' special emissary might try to repair Anglo-Nepalese relations, he received a rebuff. The Gurkhas were suspicious of British solicitude for Tibet and refused to commit military assistance. In the words of D.R. Regmi, 'it delayed British infiltration into Nepal for at least another fifty years.'[5]

Kirkpatrick, however, had made a clumsy survey of the trade routes in Nepal and observed that 'the woollen staples of Great Britain and warm flannels of the finer sort' had unlimited prospect in trans-Himalayan commerce through Nepal. In his memorandum, he urged: 'If we wish to push our commercial speculation into the western parts of Tibet .... we must for this purpose turn our eyes towards Nepal ....' Nepalease hostility is ascribed by him to 'unremitted jealousy with which the administration had all times discouraged the free ingress of strangers.'[6]

Kirkpatrick was followed by Abdul Kadir in 1795 who found Nepalese brokers between Indian merchants and Tibetan counterparts doing brisk business. The trade pact of 1792 would have undersold them. hence, the Burbar kept it inoperative. He suggested direct trade with China and Tibet bypassing Nepal. Next explorer was Captain Knox who in the wake of the Treaty of 1891 went to the Nepal Durbar with instructions to obtain information about the valuable mineral and agricultural products of Nepal, the possibility of introducing European and Indian manufacture there and opening trade with Tibet and Bhutan.

Walter Hamilton in his Gazetteer noted how European copper was driving Nepal copper out of market and the jealousy and ignorance of Nepal administration stood in the way of an extensive traffic between Tibet and the British territories.[7]

The above sketch of calculated commercial penetration should explain why the Gurkhas were wary of British intentions. Nepal suffered from anxiety and not jealousy and ignorance. She was studiedly undermined. As one sober British surveyor, Dr. Francis Buchanan-Hamilton was to point out candidly, 'I am indeed persuaded from having seen a good deal of people in this state of society that all interference of other Governments by treaty does injury and that all the merchants will suffer less and push the trade farther, if left entirely to their own exertions.'[8]

It was no wonder, therefore, when Bhim Sen Thapa (1806-87) determined to build an Asian Front against the growing British colonialism.[9] Add to that the candid remarks of Brain Hodgson, assistant resident at the Durbar on the commerce of Nepal in 1831-32 : 'It is scarcely necessary for me to remark that a connection was originally sought by us purely for commercial purposes, which purposes the Government up to the beginning of this century directly and strenuously exerted itself by arms and by diplomacy to promote'.[10]

The background of the Anglo-Nepalese War of 1814-16 can now be seen in better light. The Gurkhas sought to break the cordon of trading outposts which subverted Nepalese trade and paid the penalty by being defeated in the hands of the British and surrendering choice areas in the western, central and eastern Himalayas in the Treaty of Sagauli (1816). British success only whetted her desire to monopolise Himalayan trade across Tibet to China. To that task, Hodgson was engaged for 20 years of his life. He visualised a commercial empire for the British spreading over from India through Nepal and Tibet to China. He 'secretly and carefully applied to some of the oldest and most respectable merchants of Kathmandu and the other chief towns of the valley for conjectural estimates of the total annual amount of imports and exports and the number and capital of the chief commercial towns of the valley'.[11]

He located '52 native and 34 Indian merchants engaged in foreign commerce both with the South and the North' and found 'that the trading capital of the former was not less than 50,18,000 nor that of the latter less than 23,05,000'. A third of these merchants were natives of the plains. 'Many native merchants of Benares have established flourishing kothees at Kathmandu; the Cashmirians of Patna have had kothees there for ages past ...... lakhs of the natives of Oude, Behar and north eastern Bengal of all ranks and conditions annually resort to Kathmandu to keep the vernal festival at Pasupatikshetra. Businessmen of Benares have more than 10 kothees at Kathmandu while Bengal merchants have abstained ...... The native merchants at Calcutta have whilst there, a hard struggle to maintain with their European rivals in trade but at Kathmandu they have no one attached to the Residency and (sic.) have no access to the country ...... but every native is free to enter Nepal at leisure.'[12]

Hodgson was not content with the Anglo-Nepalese trade which averaged about 26 lakhs sicca rupees. He sketched a route from Calcutta to Pekin of 2880 miles in about 87/99 stages at the end of which 'the merchant enters Setchwan, the commercial province of China onto Yengtze and Hoangho where he may sell his European and Indian products and purchase tea or silk or other products of China'. Hodgson was the avant-gards of British colonialism in Asia, ready to move the most insurmountable mountain in the world. The schemer writes : 'But England and China and not Calcutta and China, it may be argued must be the sites of the production any consumption of the truly valuable articles of commerce of which the Nepalese and Indians would have little more than the carrying trade ... Let the native merchant of Calcutta and of Nepal ..... take up this commerce and whilst we, though not the immediate movers, shall yet reap the great advantage of it.'[13]

Hodgson's exhortation to Bengalee merchants was mainly to use them as the errand-boys of colonialism and not to foster them.

Hodgson saw Tibet as a happy hunting ground for British commerce. 'With her musk, her rhubarb, her borax, her splendid wools, her mineral and animal wealth, her universal need of good woollens and her incapacity to provide herself ot to obtain supplies from any of her neighbours, Tibet may well be believed capable of maintaining a large and valuable exchange of commodities with Great Britain through the medium of our Indian subjects and the people of Nepal.' Hodgson showed the way to intra-Asian country trade through the hills which was the main line of profit for the British in India. He even chalked out the detail of the route, manners, cost of carriage and the nature and amount of duties levied by the Nepal Government en route to Lhasa. Reviewing trade through Nepal, Hodgson appraises the Government of the enormous demand for English broad cloth in Tibet, popularity of British imitations of Indian handkerchief and cashmere shawls among the middle and lower orders in Nepal, use of English chintzes and finer cotton fabrics among upper and middle classes.

He further informs that satins, silks and velvets of China were giving way before British board cloths and velvets. Indigo and opium were in great demand in Tibet. There was demand also for British guns, scissors, mirrors, wall shades, chandeliers, tumblers, wine glasses. The people in Nepal were beginning to use British crockery and glassware at their tables. The Tibetans already used British plates and dishes.[14]

Of the exports from the hills, major items were gold of the value of Rs. 175,000 rusk (Rs. 55,000), copper Pice (2,50,000), Nepal Rs. (2,00,000), elephants (67,200), horses (45,000), timber (3,00,000) .....About paper, Hodgson writes : 'the paper of Nepal is an admirable article which ought to be substitued for flimsy paper of the plains for the records of every office in Bengal.' 'He highlights tea to be had from Lhasa 'in any quantity'. 'The Nepalese saul forest is an inexhaustible mine of timber', writes Hodgson. Copper coins were supplied in bulk from Nepal to the plains. There were lead, sulphur and zinc mines in Nepal but no skill to work them profitably. The net produce of rice of the low lands of Nepal was exported to Patna chiefly on Government account which paid for luxuries of British India imported by Nepal. Still the hills had adverse balance of trade against British India.[15] This was so by over 4 lakhs as a result of British economic penetration.[16]

It is interesting to note that from the Table illustrating Trade of Nepal that huge quantities of indigenous goods were sold by Indian merchants in Nepal. Country chintzes from Tanda, Bareilly, Furrakabad, Dacca, Jamdani, Muslins of Bihar, Santipur, Chandrakona, Sahans of Dacca, Santipur, Kora cloth of Darbhanga, Jahanabad, etc., Dacca turban, Maldah cloth, Murshidabad cloth, Benares Kinkhabs, Cashmere shawls, Bengal silk thread, pan of Bihar and Bengal, tagee horses and buffaloes were regular items of trade and accounted for more than half of the import trade. This had a special significance for the plains economy. Overland trade was one of the major outlets through which the so-called subsistence economy responded to market forces from pre-British times. The economy was sufficiently monetised and the mercantilist policies of the Company could not prevent the commercial operations of the Indian merchants through the hills. Deindustrialisation was, therefore, less complete than supposed and commercialisation of an agrarian economy was not all due to British thrust to find investments for Europe. It had grown spontaneously and created elbow-room for the Indian merchants to grow and run a parallel economy with the British.

British economic penetration to Nepal, however, meant adverse balance of trade for her, underselling of her merchants and decline of local industries like copper, iron and other indigenous wares. Worst of all, surveillance and subversion of the Durbar by the British Resident at Kathmandu, attempts to use Nepal as a transit station for trade with China and Tibet to the exclusion of Nepal, repeated trade missions pressuring her to accept commercial treaties under veiled threat compromised Nepal's independence.

The way Hodgson tried to thrust the treaty of 1835 on Nepal proved irksome to Bhim Sen Thapa. He was overzealous and peremtory in his demand for a bilateral treaty. But the Durbar was unwilling to co-operate. To the British, it was Nepalese obstinacy and ignorance. But it was Nepal's business which the British could not dictate.[17]

Nepal remained under the British umbrella throughout the period under review. The rise of Rana policy in Nepal and prompt British recognition of Jang Bahadur (1846-77) as the prime minister turned the scale in favour of the British. Jang Bahadur helped the British during the Sikh War (1848-49) and the Revolt of 1857. One of his reasons as he stated in his Despatch to the Government of India was in his own words 'that I know the power of the British Government and that put against it, although I might have temporary success for a time, my country would afterwards be ruined and the Gurkha dynasty annihilated.'[18]

We have taken a good look at Hodgson to understand the nature of British economic penetration into Nepal. But his reports remain inexhaustible for other allied questions. Hodgson in another paper on colonisation of the Himalayas urged the Government to take up colonisation of the Hills for a numbr of shrewd reasons. He writes : 'I trust .... that the general subject of the high capabilities of the climate and soil of the Himalayas and their eminent fitness for European colonisation having once been taken up will never be dropped till colonisation is a fait accompli and that the accomplishment of this greatest, surest and soundest and simplest of all political measures for the stabilisation of the British power in India may adorn the annals of Lord Canning's administration.' His main idea was to use the Hills as the European habitat nearing home weather and as the headquarter for 'drawing forth the natural resources of the region like timber, drugs, dyes, hides, horns, ghee and silk.' The European colony would be a godsend to the striving peasantry of Ireland and Scotch highlands and they would serve as a 'far more durable, safe and cheap barrier against Russian aggression than expenses of war against Russia in Persia.

The value of tea as a cash crop in the Himalayas was noticed by Hodgson and for 25 years, it was nurtured in the Residency garden into 'an indigenous and most abundant species' by him. Seeds were procured from China and planted in Kathmandu. These broad hints from Hodgson were well taken by Campbell who began developing Darjeeling into such an European settlement during his residency. Darjeeling was Sikkim territory occupied by the Gurkhas and returned to her by the British after the War of 1816 following Gurkha expulsion from that hill. In 1885, it was taken back from Sikkim to build a British sanatorium to make Hodgson's dream a reality. A compensation was paid to the Sikkim Raja. This was the beginning of the subversion of Sikkim. Because of frequent trespasses into Sikkim by Campbell, Superintendent of Darjeeling and Hooker the Botanist at the Residency, Campbell was imprisoned by the Sikkim Durbar. A military expedition was sent against the Raja to release Campbell in 1850. As a penalty, the revenue paid for Darjeeling (Rs. 50,000) was stopped and Sikkim was further deprived of the Lower Morung territory bearing Rs. 36,000 as revenue and a portion of the Hill. Thus, Sikkim was badly fleeced for questioning her stealthy subversion. The Raja was reduced to abject poverty. As Welby Jackson in his report on Darjeeling noted in 1854, 'the political relations with the Sikhimis have been much contracted as well as facilitated by the appropriation of the Morung tract by the British Government. In consequence of his outrage on the representative of the Government, the Raja is now almost without revenues and is a supplicant to the British Government for a pension'.[19] This is an example of ruthless frankness.

The petition of the Raja is a telling document. He prays for restoration of his land and compensation and assures good conduct. He begs to be excused for the ill behaviour of his father's servants and craves generosity of the Company's Government.[20]

Campbell's comment on the petition exposes the British motives behind severity of penalty on the Sikhmi Raja whose whole revenue was confiscated. Campbell writes : 'Freedom of travel in Sikkim now in abeyance and the trade with Tibet to be relieved from all exactions and hindrances in transit would greatly improve the resources of Darjeeling and add to its attractions as a sanatorium. If the Raja would agree to both and satisfy us ....; the generosity of Government might then be exhibited by a small annual gratuity to the present Raja as long as his good conduct deserved it.' Welby Jackson offers the gloss on it in his approval of the suggestion to grant pension if free travel and protection to all persons travelling under the care of British Government were conceded for private adventures.[21] He continues : 'the present disturbed state of the Chinese Government affords a good opportunity of breaking through the obstacles it has hitherto thrown in the way of travellers and of merchants bringing their goods from Tibet into the British dominion .... Free access into Tibet through the Sikkim territories would seem to be the best mode of effecting this. The road is well known from Darjeeling to Lhasa..... the distance is estimated at a month's journey and the two large towns, Phari with a population of 4000 and Geanchii Shubur with a population of 20,000 lie on the road. The value of imports into the British dominions by this route is now estimated at Rs. 50,000 annually and no doubt it might be greatly increased were greater facilities offered to the merchants'.[22]

The intransigence of the Nepal Government stood in the way of converting Nepal into an European mart and the centre of transit route to Tibet and China. Darjeeling to Lhasa through Sikkim provided an alternative route. So, Sikkim was systematically subverted on one pretext or the other especially when China was in distress around the fifties of the nineteenth century.

So strong was the commercial greed that the British did not give up lower and the upper Morung even after their acquisition of the trade route to Lhasa. Cotton, rice, hemp, jute etc. were commercially produced in these areas for trans-Himalayan and plains trade. Cotton from American seed was especially promoted. It was expected to raise production from 3,000 maunds to 60,000 maunds.[23] Darjeeling acquired a new importance in the changed circumstances. Hodgson was vindicated. It was rapidly organized as a commercial headquarter for Himalayan trade and European health resort. The route to Lhasa was mapped out by Campbell.[24]

Tea cultivation was avidly taken up. Campbell saw to its culture in the Superintendent's nurseries. Upwards of 2,000 plants were growing from 7,00 to 2,000 ft. altitude of different ages from 12 years to seedlings of a few months. He sought the expertise of Mr. Fortune on deputation to China by the Court of Directors during his return journey to Calcutta for better culture of the plant in the hills. The hill-station was steadily built up from the 1860's under the energetic J.W.Edgar, Deputy Commissioner, Darjeeling and C.E.R. Girdlestone, Resident of Nepal. New serpentine roads encircling terraced residential areas came with amazing rapidity through local funding. The queen of the hill stations was thus a byproduct of colonial exigency. The Darjeeling-Lhasa route through Sikkim in lieu of Nepal route which lay in abeyance because of 'rigid facts of geography and variable facts of politics' made Darjeeling a reality from the brainchild she was of Brain Hodgson.[25] The Anglo-Sikkimese treaty of 28th March, 1861 was the pinnacle of British success in economic penetration into the north eastern hills. Article VIII abolished all restrictions on travellers and monopolies in trade between the British territories and Sikkim. A free reciprocal intercourse and full liberty of commerce was assured. This was the euphemism for the complete subversion of Sikkim by the British.

References

1. P. Spear, *An History of India*, Vol. 2, (Pelican edition) 1973, London, p. 565; Also, P.E. Roberts, *History of British India*, 3rd edition, London, 1953, p. 280.

2. K.C. Chaudhuri, *Anglo-Nepalese relations*. Calcutta, 1960, pp. 13-33; Also, D.R. Regmi, *Modern Nepal*, Calcutta,1961, pp. 69-76.

3. N.K. Sinha, *Economic History of Bengal*, Vol. III, Calcutta,1970, pp. 46-50.

4. C.R. Markham, *Narrative of the Mission of George Bogle to Tibet etc.* London, 1876, p. 205.

5. D.R. Ragmi, *Modern Nepal*, p. 160.

6. *Kirkpatrick* quoted in Jahar Sen, *Anglo-Nepal trade in the nineteenth century*, Calcutta, 1977, pp. 23-24.

7. W. Hamilton, *East India Gazetter*, London, 1815, p. 607.

8. Jahar Sen, *Anglo-Nepal Trade*, p. 26.

9. G. C. Sastri, *Historical Glimpses of Modern Nepal*, Kathmandu n.d. pp. 25-38; Also J. Sen p. 6.

10. Brian Hodgson, Papers relative to the Colonization, Commerce, Physical Geography, Science and Culture of the Himalaya Mountains and Nepal : Selections from the Records of the Govt. of Bengal, No. xxvii, Calcutta, 1857, No. ii, Papers relative to the Commerce of Nepal, 1831-32, p. 10.

11. *Ibid.*, p. 13.

12. *Ibid.*, p. 20.

13. *Ibid.*, pp. 16-17.

14. *Ibid.*, pp. 18-30.

15. *Ibid.*, pp. 31-39.

16. *Ibid.*, pp. 14 & 47.

17. Jahar Sen, *Anglo-Nepal Trade*, p. 33.

18. *Ibid.*, p. 8.

19. Hodgson, Paper No. 1, On the Colonization of the Himalaya by Europeans in infra 10.

20. Welby Jackson, Report on Darjeeling, being No. xvii of Selections from the Records of the Bengal Govt. Calcutta, 1354, p. 3.

21. *Ibid.*, Appendix I.

22. *Ibid.*, p. 3.

23. *Ibid.*, p. 4.

24. *Ibid.*, & Campbell, Papers on Sikkim Morung, Calcutta, 1351, as in infra, 20, pp. 7-8.

25. *Ibid.*, Jackson, Appendix II.

13

The Problem of Peasantisation and Integration of the Santals in the Plains Bengal 1790-1910

The original home of the Santals in their own legend and history was at Ahuri-Pipri in the district of Hazaribag, Bihar. They began to migrate from their ancestral home to the neighbouring districts when land became inadequate for their growing human family. Or was it their traditional Jhum cultivation i.e. slash and burn method of clearing jungles and land reclamation or their ranging habit which made them wanderfolkers? Both were perhaps responsible. The santals essentially belonged to a mixed economy of shifting cultivation and forest products for sustenance. Their migrations were from forest to peripheral agriculture and back to forest when Jhum cultivation left the fields barren.[1] It was indeed a long trak from Hazaribag to Chhotanagpur, Rajmahal hills to Jungle Terry and the Junglemahals of Bengal. There is a theory held by Rev. Skrefsrud that they hailed from Samantabhum or Saont (modern Silda) in Midnapore district or Chhatna in Bankura district and then fanned out to Mayurbhanj, Bhagalpur and Chhotanagpur. It may well be that there was a criss-cross movement once they reached the Bengal plains. But both McAlpin and McPherson tend to disagree.[2] From their Hazaribag abode, they came to Chhotonagpur and Rajmahal hills including Chai-Champa (Bhagalpore foothills) and there followed an internecine war between the Santals and Malpaharias who would not yield an inch of ground to the outsiders. Malpaharias were basically averse to settled cultivation and depended on hunting and forest scrounging in the jungles skirting the Rajmahal hills.[3] The Santals failed to hold their ground and being more adaptive moved to new pastures. From the Rajmahals and Manbhum, they descended on the plains of Bengal proper. As they passed through the territories of tribal chiefs, they agreed to pay their tributes, rent and other dues to them, be it the Raja of Panchet or Pakur, Chhatna or Bishnupur, or chieftains of Midnapore. They faced their problem of integration into those pre-British tribal agrarian systems as much as the Khonds of Kalahandi or Keonjhar had faced in Orissa with the Chodagangas and Bhanjas or the Mundas, Oraons with the Chero or Nagbangshi chieftains in Bihar.[4] They were never free in absolute terms but paid a token rent or tribute to those Rajas. They were allergic to the payment of customary rent and laid a claim on the offals of the forest like the sal leaves and firewood.

Permanent Settlement and After

The British brought their own system of land revenue and cash nexus to bear on the peripheral area of Jungle terry and Junglemahals. The area was as the name conveys, a forest area yet to be brought under the plough. The Santhals were found invaluable for pioneering activities. They too needed virgin area as their old settlements had become overcrowded. they took up the challenge in the new territory. They swarmed into what came to be known as the Damin or Damin-i-koh in the Bhagalpur district. Since the time of Augustus Cleveland, Collector of Bhagalpur in 1779, the Company's Government had taken an initiative in tackling the Paharia marauders in the area and reclamation of the forest. Cleveland was a capable and astute officer. He tried his best through coercion and conciliation to get the Paharias reconciled to the British system. But they were not to be tamed into a settled life of vegetation. They preferred to range in the Rajmahal hills steeped in their hunting and gathering economy. He took their village headmen into confidence, made cash awards and recruited crack fighters among them into the Company's army but failed to woo them. The Santals who were struggling for existence in the hills against the more ferocious Paharias welcomed the offer to open up cultivation in the Damin. By 1790, the revolt of the Paharias and the Ghatwals of Birbhum was over and Pax Britannica was established by military action.[5] By 1809, the Damin area saw proliferating Santal settlements. The Santals entered the Dumka and Godda subdivisions. By 1832-33, J.P. Ward and Captain Tanner surveyed the area and put pillars of demarcation. Out of the 1366.01 square mile of damin, only 500 sq. miles were plains area comprising of 246 sq. miles of jungle and 254 sq. miles of arable land. Nevertheless, it was a big draw for the Santals and they migrated from the adjacent areas of Cuttack, Dhalbhum, Manbhum, Chhotanagpur, Hazaribag, Midnapore, Bankura and Birbhum to this new abode. The Santal adaptability has been vouchsafed by Robert Carstairs, in his classic, Harma's Village. He is all praise for their pioneering skill, agricultural application in raising rice, maize and knowledge of the forest produce, cleanliness and impeccable housekeeping. By 1851, there were 1437 villages and a total Santal population of 82795 people.[6]

J. Pontet, a non-official European was appointed the Superintendent of Damin in 1837. He took possession of the midlands and began settlement operations. In 1837-38, the revenue yielded was only Rs. 6402. It rose to Rs. 43724 in 1849-50. The Santhal population rose according to the Census of 1872 to 1914. Much of this infiltration to Damin can be accounted for by the exodus from the Jungle Mahals yet to recover from the Famine of 1770, disturbances of the 1790s, the Chuar to Bhumij Uprisings and the periodic droughts, especially between 1851-65. The Mahals were being made Regulation Districts from 1800 (Reg. X), Regulation XIII of 1833 and by 1879, the process was complete. The Santhals were granted Pattas or leases through their Majhis or headmen in the Damin whereas this was not so in the Jungle Mahals. The Santhals insisted on having survey and pattas to which the Jungle zamindars were hostile according to Gastrell who surveyed Bankura in the 60s. The Jungle Mahals as regulation districts fell into the cash nexus of high rent and rapacity of the moneylenders who charged upto 75 per cent on the loan. Even village zamindars became chronic debtors and the santhal settlers were reduced to the state of bonded labour (kamiotee). All these factors accounted for the rush to Damin in the first fifty years of the 19th century. The Damin was once again a Paradise Lost. Discontent brewed since 1830s as it was brought under Reg. 1 of 1827 in cases of dispute between the Santhals and the Mahajans by taking such cases to the Munsiff's Courts and the Superintendent of Damin had no hand in it. Bengali traders and zamindars also began to poach on it since then.[7]

From the above it is clear that the Santhals were from the very begining migratory in character moving from Hazaribag home to the Jungle Mahals and this had locked them in combat with the Malpaharias over the control of the Rajmahal hills which the latter did not yield finally to the Santhals. They moved and settled in the three jungle districts of Bankura, Midnapore and Birbhum for the early period of reclamation and then again criss-crossed to Deoghar, and Damin-I-Koh, especially its Godda region. Much of their migration can be accounted for by their reclamation and primary cultivation methods and aversion to pay regular rent except tribute. The moment this was insisted upon in any area of their domicile, they would strike the tent and move to some other forest area.[8]

This mobility, stress and migration are replete in their own songs and ballads.

(i) O you two boys on the mountain

Do not split faggots with axea

O you two girls
Do not make bundles of leaves
On the lower hill
For the tears will fall
Will fall from the eyes

(ii) Two boys, ploughing, ploughing
Ploughing on the hillside
Two girls drawing water
Water on the lower hill.
Did you see my father
O you two boys
Ploughing on the hillside?
Did you see my mother
O you two girls
Drawing water on the hill?
We saw your father
High up on the mountain
We saw your mother
Far away below.

(iii) O my love

The dust rises on the land

The sky is hot and red

Take me to another land

With shoes for the hot earth

And for the sky an umbrella

Take me to another land

On a bullock with a long and

Tufted tail

O my love

Load up some rice

take me to another land.[9]

Except for the hordes who went to the Damin, the rest of the folks decided to settle down as peasants in the forest belt of Bankura, Birbhum and Midnapore. But forests are not permanent assets unless recycled and relaid by way of social forestry. The Santhals were good at land reclamation from forest and jhoom cultivation in those lands but not so versatile in forest conservation. In the early decades of the 19th century, the Jungle Mahals were being gradually denuded of forest resources. Tertiary lands had been added to the settled plain but by the scotched earth method, the land had gone dry and barren in many places. There were few trees left to hold the soft soil. The British collectors and surveyors noticed it in their reports. The result was an agrarian economy based on crop rent or sanja which was different from barga or sharecropping developed later. The dry zone was not compensated by any complementary irrigation system or afforestation scheme by the British authorities except for bunds or tanks dug by the tribal rajas. The area was arid and the crop-yield was poor. In fact it was famine-prone and a mild drought would result in desertion and death. Robertson reports that the Santhal settlers of the district began deserting to new pastures of the lower Bengal and the Barind or north West Bengal, where there was a chronic shortage of labour for clearing of jungles. This explains how Santhals fanned out to the four corners of Bengal in search of livelihood.[10] But it is not true that they had no sense of afforestation or social forestry. Their ballads bear testimony:

(i) In the upper village

A bar tree formed her garland

In the lower village

a pipal made her necklace .....

(ii) The earth and trees

Are cut with mattock and axe

But who will stand beneath his tree?

Village headman, you are my

father

Let the jaher tree stand up again

etc.[11]

It is well known that the sal, mahua, tamarind and mango are some of the household trees of the Santhals which they recycle for domestic use. It is the wild forest they are prone to destroy without any knowledge of the ecological imbalance they might cause. There was obviously a communication gap between the British and the Santhal autonomous and mutually exclusive forest policies which should have found a via media but the course of events went otherwise.

The Problem of Peasantisation

A time came in the middle of the 19th century when there was very little open and wild forest to be cleared. This was evidently after the introduction of the forest conservation policy of the Government transplanted from British Burma. This was not just born out of greed to pry out forest products for profit. Some early British administrators were aware of the need to preserve the forest for ecological balance and agricultural sustenance. The insensate felling of trees had demonstrably balded many green area without recompense and the forest department thought of optimising forest resources by conservation, diversification and rotation of trees and saplings. But the policy, however enlightened, clashed with the survival instinct of the denizens of the forest who had come to depend on the mixed economy of poor agriculture and forest scrounging in the darkly area of frontier Bengal. While the British authorities ruthlessly pursued their forest policy of denying the needful to the Santhal community, the latter had to come to grips with the social reality of adjusting within the peasant economy of the plains. There was no longer the unclaimed forest to fall back upon except[12] for firewood, grassroots and Kendu or Sal leaves. They had to adapt themselves within the plains economy and its requirements. This would entail payment of land revenue according to the British schedule based on cash rent or commodity to be converted into cash and adjustment of the Santhal communal identity within the Bengal vertical village societies. The system of money-rent as it was done in the reclaimed areas overtook the old system of sanja which was basically payment according to capacity. The new settlement was unfair to the pioneers too exacting in its schedule of payment with which the Santhals were unaccustomed and unable to cope. As in the case of the old rajas at the initial stage of the Permanent Settlement, so also the santhals in this difficult period of integration and adjustment became dependent on the new class of moneylenders who made it their business to pay the revenue to the Government on their behalf and claimed it back with ruinous interest on it. The new order of things had an exactitude stupefying for the erstwhile carefree Santhals. They got hopelesly enmeshed in their snares. Money supply under the raj was so tight that the Santhal had to surrender the entire fruit of their labour to pay the rent an interest. It made the rent-burden enormous and the debt-burden a death-trap. The moneylender appeared as the scourage of the countryside and yet a necessary evil in the absence of modern cooperative credit societies and rural banks.[13]

Impact

What was the net impact of the new network of British administration on the Santhal Community? McAlpin maps out the peasantised Santhals in the outlying districts of Bankura, Birbhum and Midnapore in graphic detail in his authoritative report. He shows how the Santhal communities with their in-built limitations splintered off, losing much of their old communal identity. He calls these mixed villages of Santhal and Bengali population or the broken villages. They not only illustrate their segmented identities but also their restructured hierarchy of majhis, parganits and deshmajhis in the pattern of the average Bengali village. (Tables) :

	Birbhum (1881-2)		
Santhal	Villages Surveyed	Intact	Broken
203	175	70	105
	Bankura (1893)		
1200	730	115	615 [14]

In many villages, Santhal majhis were no longer existent. Villages were not just broken. They were socially fractured and intermixed. Nucleated Santhal holdings became a common sight.

The following table demonstrates the mixed population in the three districts of frontier Bengal:

1909

District	*Total Population*	*Santhals*
Birbhum	902280	47221
Bankura	1,1161411	105,682
Midnapore	2,789114	148251

Others included Bhumijs and Bengalis. Even in Silda, an ancient seat of the Santhals in Midnapore, the Santhal villages were surrounded by the Dikku villages. This was the first step in the absorption of the Santhals in the peasant society of the plains. The next step was the broken village where the Majhi or the Santhal headman was not to be found and the village was settled through a Mustagir or a new headman or the purchaser of the estate in a new settlement who tried to keep the Santhal village in tact but was not always willing to do so. The last step was a rayati settlement with the Santhal cultivators individually and finally as adhiars or mere agricultural labourers. This was the worst form or a broken village. McAlpin reports in 1909 on the basis of various district surveys undertaken from 1872 to 1893 that the broken village was a common sight in the three districts. The Santhals had not only become peasantised but also pauperised. This account was latercontradicted by K.P. Chattopadhya in a report on the Santhals in 1945-46. But there is the possibility of regrouping of the Santhals after the cadastrol surveys were conducted and favourable settlements were made with the Santhals whenever possible not only in Bengal but also in Chhotanagpur.[15]

To note certain features of their internal history, the Santhal community was not so horizontal as has been taken for granted in serious histories of the Santhals. In the most elementary differentiation, they have the Kiskus as kings, Murmus as priests, Sorens as warriors, Hembroms as the nibility, Mundis as moneylenders, Tudus as smiths and Baskes as traders. All others were mere peasants. Though this is casually dismissed as national division of labour, any careful observer will find out that the hierarchy existed in the real sense though it was not as oppressive as in a caste hierarchy.[16]

A bureaucratic hierarchy is also well known among the Santhals. It is this : The manjhi is the headman of the village. He represents the village and conducts its relations with the outsiders. The paramanick is his assistant; the Jogmajhi is the master of ceremonies in marriage etc. also a moral vensor; the Jog-Paramanik or Paranik is a deputy of the Jogmajhi; the Naiki is the village priest; the Kudum Naiki is his deputy. The Gorait or the Kontal is a peon.

Above the majhi, there is a Parganait, generally the representative of the leader who founded the village in a locality. He is assisted by a Desh-majhi who looks after the social questions, caste etc. and has the right to levy fines in the form of a supply of rice wines. It is tacit in the communal system that each official would have the privilege of paying less rent, sometime a remission upto 7 per cent.[17] The Majhi had both privileges and responsibilities. This is also evident in the following couplet:

Midnight

And the majhi's servant

The paranaiks maid

Have run away together

Sound the drums and

Blow the trumpets.[18]

This was surely not an archetypal egalitarian society. The differentiation has been highlighted because the ideal horizontal community was only a millenial dream and is often exaggerated beyond proportion by scholars and activists alike.

But the basic point remains. The Santhals were generally peace-loving, adaptive communities. They would not flare up unless strong provocations were given by the Dikku zamindars and mahajans trying to grab their improved land at a quit-rent for better settlement. If the settlement was made through the Majhi or in other words, if the decision-making was left to the Majhi for settlement of rent and thereby the village remained in tact, the Santhals were unlikely to rise in revolt.

Resistance to British Rule

The Santhal 'hool' or the Uprising of 1855 was the beginning of it all. They were in the throes of a painful integration into a new system. The uprising has been exhaustively studied by a number of historians and anthropologists. It is not within the purview of this paper. The point to draw from the Santhal Uprising is that the Santhal failed to understand why they should have to pay, more rent then mere tribute, submit to the unfavourable justice and punishment of the alien courts and denied access to the forest resources. The plains economy had its exactitude and frequent miscarriage of justice to which the Dikku peasant was also subject. It was a complex system of land market, sale, mortgage and transfer, in sum, a system of profit and exploitation. Forest resources were being denuded by the Santhals and the Government wanted to preserve them and get revenue from them. It was verily the problem of integration. The British attempt at creating a money economy out of a hopelessly tangled feudal society created such creatures of the raj as the new zamindars and jotedars and the mahajans who tried to maximise their income from the estates and tenancies after what they had to make over to the Government. They themselves were also greedy, grasping and corrupt. Once the new order was on, they mastered its intricacies and loopholes and went in a body to the peripheral districts to rob the Santhal comunity of the unearned increment of the Jungle Mahals by hook or by crook. The latter complained of the rapacity of the mahajans, their ruthlessness in reducing the community into kamioti or bonded labour, the unparalleled corruption and extortion of the police darogahs, impossibility of getting redress in the courts of formal justice which involved 50 miles of walking from the village and writing of plaint and feeing the vakils. Sido and Kano had a divine inspiration (Christian?) in their village of Barhait, broke the Sal branches and passed the message of revolt through this symbol and the 'hool' began. It was a hit and run warfare but the British had to call the army in companies and divisions to quell it finally. But conciliation followed coercion after the victory.[19]

A new non-Regulation district now called Santhal Parganas was created 1855 with special privileges of paying customary rent which was much less than rent prevailing in the plains and non-transferability of holdings. The privileges were codified in the subsequent regulations of 1872 and 1886. The Santhals in the plains were welcome to migrate to this exclusive area for them. The Santhals were also realistic. They settled down within the new agrarian framework with their keen sense of survival and revival. The forest Conservancy laws were also relaxed to some extent from 1865 to 1872 whereby the Santhals were allowed to take Sal shrubs and rollers but not to fell trees. The unjust Rent Act was also revised by 1869 and more adequately in 1885 protecting tenancy rights. But the santhal Parganas could not accomodate the proliferating community. So by 1872, they learnt to adjust to the rules and regulations of the plains of West Bengal or migrate to other districts in the Barind of North West Bengal.[20]

The majhi movements that we see after 1860 are not the traditional resistance movements like the great 'hool' but 'secondary' or 'participatory' in character. They became accustomed to the Census and settlement operations. In fact, they became attached to these operations which granted them recognition and right. The later Santhal agitations were mostly over the rent question where their guaranteed rights were being whittled down by the vested groups of Jotedars and mahajans. As O'Malley points out, Durga Majhi of Thakurpara and Brojo Majhi of Salunga, Bankura in 1906 were parganaits representing a large community of Santhals in the three districts. They had petitioned the Government for redress in rent questions failing which they were to appeal for an agitation. The babajis, half-Christians and Majhis emerged as leaders in the wake of settlement and census operations for defence of old rights and acquisition of new rights. They looked upto the Government for justice against entente of Zamindars and Mahajans. This was therefore secondary and participatory and not a millenial anti-British movement as such but had portents of such a revolution if not tackled with understanding. The movement, however, had its strength and the Government had to settle it by the offer of larger amount of concessions between 1855 and 1908. In its cultural content, the movements were marked by syncretism, absorbing elements of Hinduism and Christian religion. This was true of the Mundas and Oraons in their Khewar and Tana Bhakat agitations. This was neither detribalisation or retribalisation. This was at best revitalisation of the communities in their segmented identities. The historians and anthropologists have tried to affiliate them to their shibboleths and paradigms of enquiry. The matter should be left to a strict biography of the concerned communities.[21]

In the foregoing, it has been hopefully made clear that the Santhals were not trying to get back to their Paradise Lost. They were very much participant in the new system and only challenged its deliberate distortions. Even regulations which were oppressive could be discussed and amended once the Majhi was taken into confidence. He had to be a participant in the decision-making which would decide the destiny of the benighted Santhals. Joint decisions could settle all outstanding problems of revenue, justice, forestry and ecology but both sides should be interested in a settlement. The Santhal obduracy in refusing rent in a settled village and sticking to tribute withered away with time. But the Government and its minions were unrelenting in their efforts to raise income.[22] As a recent forest official has perceptively written, the whole of West Bengal has only 13.4% forest patch in a situation of decline and degradation. The Sal tree cannot be felled at will nor can the Santhals be kept in chains. The total concept of development must be infused among the tribal community through village Protection Committee which now number over 6/7 hundred. There must be a dialogue before action is taken. After all, the Santhals are not habitual destroyers of forest.[23] They also know how to recycle forest. The following song reveals it:

Let us plant mango and

Palm trees my brother.

Let us enclose a tank

my brother that we may be

remembered.

And on the top of the embankment

and on the Kadam tree

May the Sasung bird sing.[24]

Let us appeal to the spirit of this industrious community. Let the bird sing.

References

1. Archer, W.G : The Hill and the Flute, London,Bradley-Birt, F. B : The story of an Upland, London, 1905.

   Man, E.G. : Sonthalia and the Sonthals, Calcutta, 1861, Reprint Delhi, 1983.

   Mukherjee, C.L. : The Santhals, Calcutta, 1962.

   Culshaw, W.J. : The Tribal Heritage, London, 1949.

2. Skrefsrud, L.O.: Rev. A History of the Bankura Mission, n.d. A Grammar of the Santhal Language, Benaras, 1873.

   Also, Olav, Hodne, L.O. Skrefsrud, Missionary and Social Reformer among the Santhals of Santal Parganas, Oslo, 1966.

   Mukherjee, C. L. *op.cit.*

   Carstairs, R. : Harma's village, London, n.d. also, Carstairs, The little world of an Indian District Offers, London, 1912, pp. 219, 262, 265 & 294.

   McAlpin, M.C. : Report on the Condition of the Santhals ...... Calcutta Mc Pherson, M : Report on the Survey and Settlement Operations of the Santhal Parganas, Calcutta, 1909.

3. Baske, D.N. : Saontal Ganasangramer Itihas (Bengali) Calcutta, 1909.

Kumar, C.H. : Rev. Santal aar Pahariolo ak Itihas quoted in Baske (Santhali).

Sherwill, Captain, : Memoirs quoted in Mc. Pherson *op.cit.*

4. Baske, D.N. *op.cit.*

B.S. Das. : Civil Rebellion in the Frontier Bengal, Calcutta, 1973.

Mukherjee C.L. *op.cit.*

Singh, K.S. : The Tribal situation in India, Simla, 1972

5. McPherson, H. *op.cit.*

McAlpin, M.C. *op.cit.*

Basu, M.N. : The Boonas of Bengal, Calcutta.

Chattopadhyay, K. P. : Report on the Santals of Nothern and Western Bengal, Calcutta, 1945-46.

6. Carstairs, R. *op.cit.*

Baske D.N. *op.cit.* Chapter II.

Hunter. W.W.: The Annals of Rural Bengal (Reprint) New Delhi, 1975. pp. 300-315.

7. Baske D.N. *op.cit.* Chapter III pp. 16-18.

Man, E.G. *op.cit.*

McAlpin, M.C. *op.cit.*

O'Malley, L. S. S.: Bengal Districts Gazetteers, Santhal Parganas, 1916.

8. Baske, D.N. *op.cit.*

McAlpin, M.C. *op.cit.*

McPherson, H. *op.cit.*

9. Archer, W. G.: More Santal Songs in Man in India, Vol. xxiv No. 3. 1944.

   Also, Archer, Santal Poetry in *Man in India*, Vol. xxiii, No. 2. 1943.

10. Robertson, F. M. : Final Report on the Survey and Settlement Operations in the District of Bankura, 1926. McPherson, H.:-Do-, Santhal Parganas 1910.

11. Archer, W.G. *op.cit.* on Santhal Parganas 1910.

12. Sen, Suchibrata, The Santals of Jungle Mahals, Calcutta, 1984, pp. 22-121.

    Man, E. G. *op.cit.*

13. *Ibid.*

14. McAlpin, M.C. *op.cit.* pp. 17-30.

15. *Ibid.*

16. Mukherjee, C.L. *op.cit.* p. 19.

17. McAlpin M.C. *op.cit.* pp. 13-16.

18. Archer, W.G. : More Santal Songs *op.cit.* p. 142.

19. Culshaw, W.L.: Santhal Rebellion in Man in India Vol. xxv, No. 1, pp. 210.

    Haram, Jugia, Harkoren Mare Hapramko Reak Katha p. 243 quoted in Baske *op.cit.*

    Deshmajhi, Chhatrai : Reak Katha quoted in Baske Chap. 1.

20. McAlpin *op.cit.* Ghuznavi, Sir A, The Forests of Bengal, Calcutta, 1935, pp. 9-20.

    McPherson *op.cit.* pp. 112-40.

21. Singh, K.S. : The Tribal Situation in India *op.cit.*

22. Mc Alpin, *op.cit.* pp. 100-130.

23. Palit, S. : The Tribals and Conservation of Forests in the Indian Forester, Vol. 116, No. 2, February, 1990.

24. Man, E.G. *op.cit.* p. 7.

25. Khan, Arati : The Santals of Bankura, 1820-1920, M.Phil. Thesis, JU, 1991.

14

Green Imperialism

Green Imperialism is the title of a fascinating work by Richard Grove on the interaction between ecology and empire.[1] Though he admits that there were colonial exploitation of the natural resources of India,[2] there was also some positive contributions made by the colonial government in the field of botany, forestry and ecology. He maintains that during colonial rule, first hand knowledge of Indian flora was gathered by British forest officials and published by them. The monumental works by Roxburgh, Wallich, Cleghorn, Brandis and Ribbentrop still remain authoritative texts on Indian botany. This was the direct outcome of British pursuit of knowledge. The colonial government was also responsible for the establishment of botanical gardens and Botanical Survey for the introduction of many commercial plants and classification of Indian flora in the Linnean system. Later when the Imperial Forest Department was started in the 1850s, the government undertook to earmark valuable forests as reserve forests which were directly managed by the forest departments. The government adopted a policy of afforestation where deforestation was taking place. In this way, an ecological balance was sought to be maintained. The colonial government had, therefore made three comtributions to Indian botany :-

(a) Compilation and dissemination of botanical knowledge.

(b) Introduction of exotic commercial plants through botanical gardens like silk, cotton, tea, coffee, opium, hemp, safflour and jute into our commercial agriculture.

(c) Preservation of prime forests of teak, sal, sissom etc. by declaring them as reserve forests and afforestation. Thus an ecological balance was created in the prevaling circumstances of indiscriminate destruction of trees by local people.

What Grove tries do is in fact an apology for colonial exploitation of India's natural resources and loss of green cover. It has become in reality a defence of green imperalism for maintaining an ecological balance in its favour. A review of British botanical and forest policy will make it clear.

The botanical gardens were set up to act as nurseries for commercial plants which, after successful experiments, could be transplanted as plantations elsewhere.

Tea, coffee, opium, jute etc. are ready examples. They were initially state monopolies and later passed on to British capitalists. They were raised, financed and controlled for export abroad by British Agency Houses. They were no profitable additions to Indian flora. The British made enormous profits out of them which caused a drain of wealth from India till Independence.[3]

On a review of the British Indian forest policy, it can be found that it was not so much guided by thirst for knowledge or striking an ecological balance. The primary motive was once again commercial cupidity. As Cleghorn admits, in the early nineteenth century, Indian prime forests were utilised for ship-building purposes. Dockyards were constructed in Calcutta, Cochin and other places where smaller vessels were built for merchant shipping. This had taken a toll of Indian forests.[4]

Then came the age of the railways since the middle of the nineteenth century. Indian forests were ruthlessly exploited for making railway sleepers not only in India but also in other parts of the empire including the metropolis. This loss of green cover could not be made up by a show of afforestation.[5]

Indian timber was also used and exported for making furniture, doors, staircases and dance floors both here and abroad. Valueable timber was supplied to British capitalists for this purpose. Forests were also leased out to them for making paper pulp for their paper mills, adhesives, wax and honey the wax. These were sheer plunders and were not compensated for by adequate afforestation.[6]

By the forest Acts of 1855, 1865[7] and 1878[8], forests were classified and some prime forests were claimed as government reserve forests. They were put under forest administration. The policy was not so much to preserve the forests as to establish a monopoly of the timber. The original tribal people who were denizens of these forests were gradually pushed out of their habitat by this imperial policy of fencing the forest. Very soon the actual owners of the forests were declared as poachers and illegal occupants. Most of these people depended on these forests for their livelihood. But under the new acts, they were not only evicted, they were not even allowed to take a bundle of fire-wood. They were frequently held and punished for such encroachments and stealing of forest products. Under the new forest administration, forest officials from the Conservator down to the Ranger were encouraged to maximise revenue[9] from the forest. Their promotion depended on this achievement. As a result, there was wanton plunder of the forest and predictably more deforestation than afforestation. Many green areas turned bald as a consequence.

The government and the British capitalists together unleashed such an agressive policy of commercial exploitation of the forests that by 1947 the green cover shrank from 60 per cent in the fifties of the nineteenth century to a mere 33.3 per cent. This extraction totally falsifies the argument that green imperialism also had a benevolent face, that it cared for ecological balance of some sort.[10]

In the end, it has to be admitted, however, that plunder apart, the colonial government was instrumental in the gathering of forest information and publication of fundamental works on Indian botany and forestry. They had also set up a well-knit and structured forest administration and the pious intention was always there to conserve forests for ecological balance. These structures remain as memorable legacies for free India. Only a real concern for ecology, environment and social forestry for the tribal people can revitalise it. But the Indian operators appear to be more mercenary and they have made a mockery of these British institutions. The green cover till now has been further reduced to a precarious 18 per cent. The wood has become now unquiet and Chipko and similar movements by the NGOs have now surfaced to combat the inroads of the forest mafia.

References

1. Richard Grove, Green Imperialism, New Delhi 1998, Introduction, pp. 1-15.

2. Ramchandra Guha-The Unquiet Woods, New Delhi, 1999.

   Guha's emphasis on the disjuncture owing to colonial rule leading to a neglect of ecological changes, of chapter on "Forest unrest and Chipko Movement."

3. Abhijit Mukhopadhyay: 'Natural Sciences in colonial context: History of the Botanical Gardens and Agri-Horticultural Society', unpublished Ph.D. Thesis, Jadavpur University, Cal., 1998.

4. Ramchandra Guha, *op.cit.*, chapter on "Forestry in British and Post-British India".

   He argues that the practices of colonial forestry were largely an outgrowth of the revenue and strategic needs of the empires.

   Also, M. Gadgil and R. Guha, 'This Fissured Land', New Delhi, 1992 quoting Cleghorn, Indian Forestry for arive to obtain timber for shipping in Malabari in 1806.

5(a) D. Brandis "Memorendum on the supply of Railway Sleepers of the Himalayan Pines Impregnated in India", Indian Forests, Vol-IV, (1879), p. 366. The buildings of the first railway line from Bombay to Thane in 1853, and the subsequent expansion of the railways, generated a high level of demand for wood. A mile of broad-gauge rail track required about 1,800 to 2,000 sleepers. By 1878 over two million sleepers had been used for the Himalayas. Sal and teak were the main sources of wood.

(b) "Railway Fuel in Punjab", Calcutta Review, Vol. XI, VI, (1867) pp-262-327, "In addition to timber, railways generated a demand for fuel, especially in areas like the Punjab where there was no coal."

(c) Whittaker's Report to DC, Betul, July, 1859.

   Whittaker focussed official attention on the situation in Betul where he was deputed to supply 200,000 logs for the Indian Peninsular Railway but found "nearly the whole of the timber iltegally cut by the Bohras, native timber dealers."

6. A.S. Rawat-A History of Forestry in India, Indus Publication, New Delhi, 1991, chapter on Forest Management.

M. Gadgil and R. Guha-This Fissured Land, OUP, New Delhi, 1992 p. 135.

It served first, during the era of railway expansion and later during the two World Wars. At the same time, the forest department had to generate an adequate revenue in keeping with a carinal principle of imperial policy, namely that the administrative machinery had to be self supporting.

7. M. Gadgil and R. Guha-This Fissured Land, OUP, New Delhi 1992, p. 123.

   The first attempt at asserting stake monopoly was through the Indian Forest Act of 1865. This was replaced thirteen years later by a far more comprehensive piece of legislation.

8. Mahesh Rangarajan, Fencing the Forest, OUP, New Delhi, 1999 (Paperback), chapter on Tribal opposition to Forest Acts.

   The 1878 Act provided the underpinnings for the 'scientific management of forests, enabling the working of compact blocks of forest for commercial timber production.

9. *Ibid.*

10. In the War of 1914-18, timber and bamboos were supplied for the construction of bridges, piers, wharves, buildings, masts and ships. In little over a year (April 1917 to October 1918) 228,076 tonnes of timber (excluding railway sleepers) were supplied by the specially created 'timber branch' of munitions board and 30,000 tonnes of fodder grass exported to help military operation in Egypt and Iraq.

    Cf. Gadgil and Guha *op.cit.* p.138.

15

Two Cheers for Environment : Tagore and Gandhi on Environmental Issues

Environment has now gripped our attention as a global issue. Loss of green cover on the face of the earth and industrial pollution have led to green-house effect on environment and endangered the lives of all living beings. All over the world, environment has become a focal point of discussion in all fora. It is no longer a parochial issue confined to a specific country. The balding of one country affects the environment of another. This has necessitated a historical evaluation of extraction of natural resources in the last century. Two father figures of South Asia figure prominently here namely Tagore and Gandhi. Apart from there literary and political preoccupation both had deeply thought and written on environmental issues. It has become urgent to recall their thoughts in the current ecological and environmental crisis.

It may be surmised that Tagore was aware of green imperialism during British rule which led to the loss of green cover from sixty-percent in 1865 to thirty percent in 1947.[1] It has been recently argued by Richard Grove that green imperialism was not marked by unalloyed commercial cupidity. Thirst for knowledge and ecological concern also actuated British forest policy.[2] But as Madhav Gadgil and Ramchandra Guha have pointed out, profit-making was the main incentive and plunder of timber caused more deforestation than afforestation. Conservation was only to pick out prime forests for exclusive exploitation. It all began in the early nineteenth century, when timber was extracted for ship building first and then for providing sleepers to the railways from the mid-fifties. There was a maximisation of forest revenue later on by sale of timber for furniture and building material. British capitalists were given the privilege to set up forest-based industries like the paper mill. There was a cartel of foresters, timber merchants and lorry contractors to syphon off valuable timber illegally at a throw-away price. Indian sub-contractors were no less ruthless in this game.[3] The overall balding and baring of sylvan surroundings of village India resulted in the embellishment of urban centres at the cost of villages. The continuous drain of resources from the villages impoverished them as they lost their natural sustenance. The two lasting legacies of the Raj were however compilation of information on Indian forestry and the iron-frame of forest administration which came to use in free India.[4]

Tagore seems to have noted everything stated above with concern and he made general comments on this natural disaster without imputing it to the Raj or its collaborators. From his numerous Bengali essays on rural reconstruction and his statements recorded by L.K. Elmhirst in his book 'Poet and Ploughman', it is clear that Tagore reseated the practice of pauperising the village to pamper cities. He warned that one day it would backfire and there would be no resources left and rural destitutes would engulf the cities.[5]

From his deep thoughts on the matter of ecological balance, he came out with some sage observations. Tagore's first recipe was to regard both anthropomorphic and non-anthropomorphic creations as divine manifestation resting on perfect complimentarity. This stress on complimentarity of nature with man was his new message to destroyers of nature.[6] From this Tagore took the next step towards the formulation of his idea of communion with nature. This was evident in the 'Halakarshan' or furrowing 'Briksharopan' or tree-planting festivals in Santiniketan. Tagore has many songs and verses for these occassions. Tagore delighted in nurturing new saplings and old trees alike. Visva Bharati became a garden University. Sheltering under the tall Chatim and Akashmani trees, Tagore has written a memorable long poem on his pet creeper Nilmanilata which he tended with motherly care. The school classes were held in the Amrakunja or the mango grove in the open air. This was the best example of communion with nature. Students felt related while learning with pleasure.[7]

Tagore wanted this natural abandon to be part and parcel of human life. For this man had to cease from relentless exploitation of nature. He quoted from Upanishad like T.S. Eliot in 'Wasteland' urging men to avoid consumerism and adopt plain living and high thinking. This was the only solution to prevent depletion of natural resources causing natural catastrophe. He condemns human greed to possess more and consume more till mother earth gasped in atrophy.[8]

This further led Tagore to condemn colonisation and imperialism by the western powers over Asia, Africa and latin America and plunder of natural resources of these unspoilt continents. This is replete in his poem 'Africa' where he condemns western imperialists for planting their teeth and claws into its body-fabric.[9]

Reduction of wants and simple living are then the two signal messages of Tagore to avoid natural revenge of a mighty holocaust. Human civilization would be better preserved in a purer form by learning to live in harmony with nature.

We may now turn to Mahatma Gandhi's thoughts on environment. Tagore was a universalist but still he addressed the question of environment from both universalist and nationalist points of view. Gandhi had an out and out nationalist point of view. Gandhi had studied law and became a barrister in London. He had been exposed to the vegaries of industrial society. He, therefore, did not want his country to become its replica. In his 'Hind Swaraj' (1910) he contrasted the eastern and western civilizations. He criticised the British industrial society as a matrialist conumer society which thrived on the competition and maximization of private gain. There was no end to plunder of nature and gorging of consumer goods. Nature was poorer for it and the symbiotic relation between man and nature got ruptured because of this predatory process. Narrow self-seeking made men crooks in their profession. Gandhi was very critical of doctors and lawyers in this regard. He even made caustic remarks about Western women who had become aggressive and given up their modesty. He wanted to see India as a contrast to the western model. Indian life of modesty, simplicity and friendly community relationship nurtured by nature was extolled by him in reply to western constructs of India.[10]

While in South Africa, he was profoundly influenced by two western thinkers, namely, Tolstoy and Ruskin both of whom believed in simple life in communion with nature. Ruskin went one step further by saying that tillers of the soil were purest as they were closest to mother earth. He lived in two agricultural farms Tolstoy and Phoenix farms in the lap of nature. He was so sensitive to exploitation of nature that he refused to use even an electrical letter press for publishing Indian opinion, his newspaper but he was prevailed upon by his 'Colleagues to give up the extreme stand. In South Africa, he lived the life of an ascetic raising own foodcrops and cleaning up his environs. Here he adopted fasting as a medicine and discipline of mind and practised naturopathy. In evolving his political-cum-moral principles, he laid stress on 'Satyagraha' or penchant for truth and 'Ahimsa' or 'Non-Violence'. The path of truth made the seeker of truth stronger than the adversary conching in lies and thereby half the battle was won. Ahimsa followed from it. The violence of the armed authorities against the disarmed seekers of justice became immoral in character and sapped the enemy's strength. There were all quintessential parts of his moral philosophy drawn from the concept of natural justice. With this harmless weapons he won the civil rights of Indians against the military regime of General Smuts.[11]

Back in India he begun his struggle for Indian independence with the same moral armament. He started his rural mobilisation in the grab of an ascetic in consonance with the natural surroundings of poor people. He went to Champaran in North Bihar and his first step in the Harijan village was to sweep the dusty roads like St. Francis of Assisi. He enquired about the dirty clothes of the Harijan women and found that they were unable to afford even two pieces of cloth for their daily wear. He instantly thought of Charka as the only remedy. This was the symbol of self help which empowered women to produce their own cloth and some extra for earning. Charka was a pollution-free cheap device which could be manufactured by poor people themselves and it formed part of the natural scenario.[12]

Gandhi has been heavily critisied for his advocacy of Charka. Even tagore blamed him for putting India backward towards primitive handicrafts. But Gandhi was grossly misunderstood. Charka was the appropiate tool for rural reconstruction. But he was basically not opposed to big business. The financial bastion of the Indian National Congress was built on the big business houses like the Birlas, Bajajs, Purushottam Das Thakurdas, Lalbhai Kusturbhai etc. He only made a fervent appeal to them to act as trustees of the people and dedicate themselves to public welfare. This principle also evolved out of his concern for natural harmony.[13] Wherever he went in India he would insist on physical cleanliness and teach by his own examples. His concern for the Harijans was not simply a political gimmick, it came out of his deep conviction that the mehter was the mother. As a mother cleaned up her baby for its divine smile, the mehter also clean up the whole society by his scavenging duties. Gandhi would go to the Bhangi colony, sit on their Charpais and scold them for their wayword habits. His paper Harijan should not be regarded as a political mouthpiece for rallying the untouchables. His quest for natural justice found expression here.[14] His 'Ramrajya' was not a communal concept, but a kingdom of justice. His Ramdhun was an eclectic song addressed to all communities. The Sabarmati Ashram was an abode of peace, harmony, love and worship of nature. Here he would not even allow the killing of a snake and refuse to wear a floral-wreath and use of wreath of thread instead. He was strictly a vegetarian and his meals were punctuated by frequent fasts.[14]

He was critisied by contemporaries and historians an a champion of 'Hindutwa' but he was assassinated by Nathuram Godse for being partial to the Muslims-during partition parleys and the final divide. His concern for undivided India can be perfectly situated in the dialogue on environment.[15]

The two personalities who constitute the panoramas of concerned views on environment frequently interacted with each other. Let us quote from a Welcome address of Tagore to Gandhi when the latter visited Santiniketan in 1940 : 'Just as this moment there are problems that darken out destiny. This, we know are crowding your path and none of us in free from their attack. Let us for a while pass beyond the bounds of this turmoil and make our meeting today a simple meeting of hearts whose memory will remain when all the moral confusions of our distracted politics will be allayed and the eternal value of our true endeavours will be revealed." Gandhi in reply to his reception on a later visit in March 1940 reciprocated this sentiment in his speech on the occassion;

"I have not come here as a stranger or a guest, Santiniketan has been more than a home to me. It was here that the members of my South African family found warm hospitality in 1914 pending my arrival from England and I too found shelter here for nearly a month. The memories of those days crowd in upon me as I see you all here assembled before me .... In a letter to a friend the other day I described my present trip to

Santiniketan and Malikanda as a pilgrimage. Santiniketan has truly this time proved for me a 'niketan' of 'shanti'-an abode of peace, I have come here leaving behind me all the cares and burdens of politics, simply to have Gurudev's darshan and blessings.[16]

In this strife-torn world of ecological imbalance these two stars of Indian firmament lighten our dark and wending path to warn us againt destruction of nature which in turn means delf-destruction.

References

1. Madhav Gadgil and Ramachandra Guha ; *This Fissured Land, An Ecological History of India*, New Delhi, 1992, cf. Colonial Period.

2. Richard H Srove; *Green Imperialism*, OUP, Delhi, 1995, p. 382.

3. M. Gadgil and R. Guha, *op.cit.* and Specially R. Guha for the Post-Colonial Period.

4. *Op.cit.*

5. L.K. Elmhirst; *The Poet and Ploughman*, Introduction, Visvabharati, 1932, also see R. Tagore, *Upekshita Palli* and *Palli Prakriti* in *Rabindra Rachanabali,* Govt. of W.B. 1961, Vol. XII, p. 528.

6. R.N. Tagore; *Essays on Santiniketan*, Visvabharati 2000, c.f. 'Topoban' and 'Viswabodh' (in Bengali).

7. R.N. Tagore; '*Santiniketan Brahmacharyashram*' in complete works of Rabindranath, British Centinary edition, Govt. of W.B., 1961, Vol. XI, p. 824.

8. R.N. Tagore; *Salyatar Samkat* (Crisis in Civilization) *Ibid.*, Vol-XIII, p. 410.

9. R. N. Tagore; *Sanchayita* (Collection of Poems) Visvabharati, 1961, c.f. 'Africa'.

10. M. K. Gandhi; Hind Swaraj or Indian Home Rule from the *Collected works of Mahatma Gandhi*, Vol-X, Delhi, 1963, Passim.

11. M.K. Gandhi; *The Story of My Experiment with Truth; An Autobiography*, London, 1949, Passim.

12. *Ibid.*

13. D. Tripathi; *Business and Politics in India*, Delhi, 1954, Passim, also see Mahadev Desai; *Day to Day with Gandhi*, Vol. IX, Varanasi, 1974, p. 171.

14. Louis Fisher; *The Life of Mahatma Gandhi*, London 1962, p. 252, Also see M. Desai; *Day to Day with Gandhi*, op.cit, p. 182.

15. S. Settar and I.B. Gupta (ed.); *Pangs of Partition*, Vol-I, Manohar, Delhi, 2002 c.f. My Article on Mahatma Gandhi and the Partition of India.

16. S. Bhattacharya (ed.); *The Mahatma and the Poet, Letters and Debates between Gandhi and Tagore*, 1915-1941, National Book Trust, New Delhi, 1997, pp. 208-209.

Index

D

E

F